CHEMICAL METHODS FOR ASSESSING BIO-AVAILABLE METALS IN SLUDGES AND SOILS

Proceedings of a Seminar organized by the Commission of the European Communities, Directorate-General Science, Research and Development, Environment Research Programme, held in Münster, Federal Republic of Germany, 11–13 April 1984

CHEMICAL METHODS FOR ASSESSING BIO-AVAILABLE METALS IN SLUDGES AND SOILS

Edited by

R. LESCHBER

Institut für Wasser-, Boden- und Lufthygiene des Bundesgesundheitsamtes, Berlin

R. D. DAVIS

Water Research Centre, Medmenham, UK

and

P. L'HERMITE

Commission of the European Communities, Brussels

ELSEVIER APPLIED SCIENCE PUBLISHERS
LONDON and NEW YORK

ELSEVIER APPLIED SCIENCE PUBLISHERS LTD
Crown House, Linton Road, Barking, Essex IG11 8JU, England

Sole Distributor in the USA and Canada
ELSEVIER SCIENCE PUBLISHING CO., INC.
52 Vanderbilt Avenue, New York, NY 10017, USA

British Library Cataloguing in Publication Data

Chemical methods for assessing bio-available
metals in sludges and soils.
1. Sewage sludge—Analysis. 2. Metals—
Analysis. 3. Soils—Metal content.
4. Soils—Analysis.
I. Leschber, R. II. Davis, R. D.
III. L'Hermite, P. IV. Commission of the
European Communities.
628'.2 TD767.4

ISBN 0-85334-359-4

WITH 33 TABLES AND 28 ILLUSTRATIONS

Publication arrangements by Commission of the European Communities, Directorate-General Information Market and Innovation, Luxembourg

EUR 9538

Printed in Great Britain by Galliard (Printers) Ltd, Great Yarmouth

FOREWORD

This publication constitutes the Proceedings of a Seminar held at the Josef-Konig Institute, Nevinghoff, Münster in the Federal Republic of Germany on April 11-13, 1984, under the auspices of the Commission of the European Communities, as part of the Concerted Action COST 681 "Treatment and Use of Organic Sludges and Liquid Agricultural Wastes". The Seminar was organized jointly by Working Parties 2 (Chemical Pollution of Sludge) and 5 (Environmental Effects) of COST 681 and its success owes much to the organisational work of Dr Gerd Crossmann of the Josef-Konig Institute.

Utilization of sewage sludge on agricultural land is an essential disposal route accounting for about 30 per cent of the $6x10^6$ tonnes (dry solids) of the sewage sludge produced annually in the EC countries. Sewage sludge contains higher concentrations of various metals than most agricultural soils and when sludge is applied to the land the metals tend to accumulate in the cultivated layer. It is essential that metal concentrations in sludge-treated soils are not allowed to reach levels which could deleteriously affect soil fertility or animal and human foodchains. For this reason there are ordinances and regulations in force in most of the countries where sludge is used on land which include guide or limit values for maximum permissible metal concentrations in sludge treated soils. These limits usually apply to the total content (often strong acid extractable concentration) of metal in the soil. For regulatory purposes this approach is both convenient and cautious since only a small proportion of the total content of metal is likely to be environmentally active in terms of its availability for crop uptake or for leaching into groundwater. Nevertheless, the total content of metals present in soil is in most cases of minor importance for calculating the extent of crop uptake and consequent phytotoxicity and entry of metals into the food chain. Also, the different forms of binding of the metals and how these are influenced by soil chemistry, plant type and climatic and environmental conditions are of major importance in the assessment of potential pollution.

These mechanisms have conventionally been studied in two different ways:

- agricultural trials with selected indicator plants to measure bioavailability to crops directly.
- simulation of heavy metal mobilization (solubilization) and potential bioavailability by physico-chemical model procedures.

Agricultural trials are time-consuming. The natural growth of the test plants - even under optimized greenhouse conditions - requires sufficient time, and standardized conditions especially in pot trials cannot be assumed to correspond with conditions in the field. Furthermore, test plants chosen for reasons of practicability may not be relevant to the human diet and estimation of food chain effects.

When applying a physico-chemical test for the determination of the mobilization of heavy metals there are difficulties in realistically and comprehensively simulating the complex soil and plant-dependent metabolic reactions involved . A great number of chemical methods for assessing bioavailability have been devised over the years but their application has been limited by analytical constraints and a lack of theoretical justification. However, recent advances in analytical techniques and a better understanding of soil chemistry have led to the development of more promising chemical methods for assessing the bioavailability of metals in sludge-treated soils.

It was the purpose of the Seminar to review these new developments in chemical methods for assessing bioavailability and to identify those most suitable for research purposes, for routine monitoring of sludge-treated soils and for improving the basis of metal limits for such soils. The papers presented dealt with a variety of new methods ranging from the use of soil solution and single-step neutral salt extractants to more complex multiple-step procedures. The papers emphasise the extractability of metals (principally cadmium, copper, nickel, zinc and lead) by the chosen procedures in relation to soil (and in one case sediment) chemistry and bioavailability to crop plants. The Proceedings include a synopsis of the discussions after each paper and also the overall conclusions.

CONTENTS

CHEMICAL FORMS AND REACTIVITIES OF METALS IN SEDIMENTS

ULRICH FÖRSTNER

Arbeitsbereich Umweltschutztechnik
Technische Universität Hamburg-Harburg
Postfach 90 14 03, D-2100 Hamburg 90, F.R.G.

Summary

Sediment analyses are used to pin-point major sources of pollutants and to estimate the toxicity potential of dredged materials on agricultural land. For source assessment standardization is needed with respect to grain size effects; various methods are described, e.g., extrapolation techniques, mineral corrections, mechanical fractionation, chemical extraction of mobile fractions of metals, comparison with conservative elements. Only part of the metals present in sediments or sludges are involved in short-term geochemical processes and/or are bioavailable. Hydrous Fe- and Mn-oxides as well as organic matter, partly as coatings or films on detrital grains, are important substrates for the interactions with dissolved metal species in aquatic systems. For the differentiation of the relative bonding strength of metals on various solid phases and for the estimation of their potential reactivity under variable environmental conditions sequential extraction procedures were used. While these determinations seem to pose basically operational problems, the correlation of the data from solid speciation with the processes and extent of the biological uptake is still unsatisfactory. This is mainly due to the competition between sorption sites on solid matter and biological processes for dissolved metals. A uniform assessment scheme for the "bioavailability" can therefore not be established on the basis of extraction chemical data from sediment samples.

1. INTRODUCTION

The major uses of sediment analyses - especially on trace metals, but also on artificial radionuclides and on persistent organic chemicals such as polycyclic aromatic and halogenated hydrocarbons can be defined as follows:

- the identification, surveillance, monitoring and control of sources and distribution of pollutants;
- the evaluation of the environmental impact of polluted sediments.

Sediments reflect the current quality of the water systems as well as the historical development of certain hydrologic and chemical parameters. Comparitive analysis of the total concentrations of longitudinal profiles and sediment cores is performed to determine metal anomalies in zones of mineralization as well as from pollution sources. The study of dated sediment cores has proved especially useful as it provides a historical record of the various influences on the aquatic system by indicating both the natural background levels and the man-induced accumulation of elements over an extended period of time. Sediment analysis is used for selecting of critical sites for routine water sampling for contaminants, that, upon being discharged to surface waters, do not remain soluble, since they are rapidly adsorbed by particulate matter and may thus escape detection by water analysis. Thus, sediment data play an increasing role within the framework of environmental forensic investigations (88), particularly in those cases, in which a short-term or past pollution event is not or only insufficiently traceable from water analysis.

In addition, sediments are increasingly recognized as a pollutant proper and as a carrier and possible source of contaminants in aquatic systems as well as with respect to the biological effects of polluted solid materials on agricultural land. Metals, for example, are not necessarily fixed permanently by the sediment, but may be recycled via biologicalal and chemical agents, both within the sedimentary compartment and also back to the water column. This is especially valid for "dredged materials", which threaten not only organisms but also the quality of water. Human activity promote the accumulation of polluted sediments in the water courses and the resulting increased maintenance dredging results in a large amount of contaminated solids for which safe disposal sites have to be found.

In the discussion of environmental problems with metals it is increasingly recognized that the ecotoxicological significance of metal inputs is determined rather by the specific form and reactivity of a metal compound than be its accumulation rate. This effect typically refers to surface-associated metal pollutants which may be primarily volatilized during high-temperature processes or become partly mobilized in the aquatic milieu by changes of the pH- and redox conditions, by increased salinities or concentrations of organic chelators (36). It is a general experience from sediment studies, that the surplus of metal contaminants introduced into the aquatic system by man's activities usually exists in relatively unstable chemical forms and should, therefore, be more accessible for short- and middle-term geochemical processes - including biological uptake - than the detrital, predominantly natural metal compounds (104).

Since adsorption of pollutants onto airborne and waterborne particles is a primary factor in determining the transport, deposition, reactivity, and potential toxicity of these materials, analytical methods should be related to the chemistry of the particle's surface and/or to the metal species highly enriched on the surface. Basically there are three methodological concepts for determining the distribution of an element within or among small particles (52,91):

(i) Analysis of single particles by X-ray fluorescence using either a scanning electron microscope (SEM) or an electron microprobe can identify differences in the matrix composition between individual particles. The total concentration of the element can be determined as a function of particle size, e.g. by sequential filtration (34). If the concentration increases as particle size decreases, it is often assumed that the element is located on the surface of the particle (34). Other physical fractionation and preconcentration methods include density and magnetic separations (28,80). Direct species determinations can by performed e.g. by X-ray diffractometry (XRD), Fourier Transform Infrared Spectroscopy (FT-IR) and Proton-Induced X-Ray Emission (PIXE) analysis.

(ii) The surface of the particles can be studied directly by the use of electron microprobe X-ray emission spectrometry (EMP), electron spectroscopy for chemical analysis (ESCA), Auger electron spectrometry (AES), and secondary ion-mass spectrometry (SIMS). Depth-profile analysis is used to determine the variation of chemical composition with depth below the original surface (75).

(iii) Solvent leaching - apart from the characterization of the reactivity of specific metals - can provide information on the behaviour of metal pollutants under typical environmental conditions. Common single reagent leachate tests, e.g. U.S.EPA (118), ASTM (5), IAEA (62), ICES (63), and

German Water Chemistry Group (Deutsche Einheitsverfahren;27) use either distilled water or acetic acid. A large number of test procedures have been designed particularly for soil studies (64); partly used organic chelators such as EDTA, DTPA, both as single extractants or in sequential procedures. A single step method of the U.S.EPA (51) designed for studies on the leachability of waste products consists of a mixture of sodium acetate, acetic acid, glycine, pyrogallol, and iron sulfate. In the sediment-petrographic field, interest was focussed initially on the differentiation between authigenic and detrital phases in ferromanganese concretions from the sea floor and in carbonate rocks (17,58). Despite differing objectives, the procedures presented here are similar to those used in earlier sediment studies. In general, they serve for the determination of the major accumulative phases for metals in particulates and the assessment of diagenetic changes. In the ore-related studies, such phase differentiation can give information as to economically favourable processing methods; in environmentally relevant investigations indications are expected as to the relative stability of the various phases to chemical transformations and to the possibility of release of metals into solutions.

2. SAMPLING, SAMPLE PROCESSING AND ANALYSIS

Sampling procedure for source reconnaissance analysis are described by Förstner & Salomons (38); sediment collection and processing for selective extraction (Section 5) has been described by Jenne et al. (67):

Surface bottom sediments are usually taken by a grab sampler of the Van Veen or Ekman-Birge type from a depth of 15-20 cm. Material from the upper, flakey, light brown, oxidized layer is generally dissimilar to the layers below it. It is suggested that the chiefly dark layers directly underneath (ca. 1-3 cm depth) are more representative of the pollution situation over the last few years, especially in river deposits exhibiting rapidly fluctuating sedimentation rates, and should be given priority for subsequent investigations. To complement this, surface sediment (current contamination) as well as samples from deeper sections (10-20 cm) should be examined. In environments with a relatively uniform sedimentation, for example in lakes and in marine coastal basins, where the deposits are fine-grained and occur at a rate of 1 to 5 mm yr^{-1}, a more favourable procedure involves the taking of vertical profiles with a piston corer. A core profile of approximately 1 m covers a historical period of at least 200 years, and its development can be traced by virtue of the metal content in the individual layers; dating of the layers, e.g. by radioactive isotopes of lead (Pb-210) and cesium (Cs-137) may be useful for such studies. Suspended sediments are recovered by filtration, continuous-flow centrifugation or by sediment traps. In rivers, water discharge rates must be known and sampling frequently repeated.

2.1. Suspended sediment

If amount of suspended sediment and associated trace elements are to be calculated, a representative suspended sediment sample must be taken by using a depth-integrating sampling procedure, for which any rubber parts of the sampling equipment are replaced by silicone ones, and exposed metal coated with a suitable plastic or replaced with Teflon (4,119). A 0.1 µm membrane will normally separate the particulate and dissolved phases effectively enough to provide valid data for thermodynamic geochemical calculations, whereas the 0.45 µm membrane may not, mainly with respect to Al, Fe and Mn (71). Metals such as Cu, Zn and Mn can be leached from filter membranes in sufficient quantities to contribute $>$1 µg/l to the initial portion of the filter if the membranes are not adequately preleached (123). A related problem that can arise during filtration os the loss of trace elements from solution due to adsorption onto the filter membranes (72). Filtration time can be reduced by using backflushing filters (73).

2.2. Bed sediments

Fewer bed sediment samples than those of water or suspended sediments may be sufficient to obtain a reliable estimate of the trace metal concentration of a river cross-section. Samples from recently deposited bed sediment should be taken from 5 or more equidistant points across the stream and from a depth of approximately 1-2 cm. Anoxic sediment samples require different sampling preservation techniques such as oxygen exclusion. Drying and freezing (also free-drying) of the samples should be avoided for material designated for extraction procedures. If total analyses or strong acid digestion is planned, the sediment is dried at 60°C, crushed and stored; for mass calculation, reweighing after drying at 105°C may become necessary.

2.3. Extraction of pore waters

Extraction of interstitial waters from sediments is performed by leaching, centrifugation or squeezing; oxidation must be prevented during these procedures. In situ methods are considered more promising, particularly for studies in the sediment-water interface. A technique described by Mayer (87) consists of placing a dialysis bag with distilled water into the sediment and allowing equilibration to take place between the inserted and ambient water over a period of some 7-8 h at 25°C. Vessels covered by dialysis membranes can be inserted in close intervals (1 cm) for the analysis of concentration gradients near the sediment surface (57).

3. GRAIN SIZE EFFECTS

Two effects must be considered when sediment analyses are used for the identification of pollution sources. Firstly, under conditions of higher water discharges erosion of the river bed takes place and generally leads to a lower degree of local contamination. (For brief periods during the initial stages of storm runoff, there is often a very significant increase in the metal concentrations in both aqueous and solid phases which can be explained by a flushing effect.) Secondly, it is imperative to base metal analyses, particularly those from river sediments, on a standardized procedure with regard to particle size, since there is a marked increase in the contents of metals as sediment particle size decreases.

3.1. Grain-size dependencies of metal concentration in sediments

In examples from the Saddle River presented in Figure 1 (after data of Wilber & Hunter /122/), the grain size dependencies of heavy metals are compared in samples taken upstream and downstream of an urbanized area at Lodi (New Jersey). In the less contaminated material (e.g. from river mile 16.6) a general decrease of metal concentrations is found with increasing particle diameter. In the polluted material there is a typical increase of metal concentrations in the clay and silt fractions; it is conceivable that the metal enrichment is due to organic matter not removed during sewage treatment processes. A characteristic feature of the polluted sediments (e.g., from river mile 0.5) is the increase of metal concentrations (except for Mn) in the medium and coarse sand fractions, particularly significant for Pb, Cu, Ni, Cr, Cd and Zn (fig. 1 right), but also evident for iron (fig. 1 left). This is probably due to the input of coarse waste particles.

The decrease of metal concentrations in the range 10μm - 1000 μm grain diameter should be even more pronounced if the mechanical fractionation

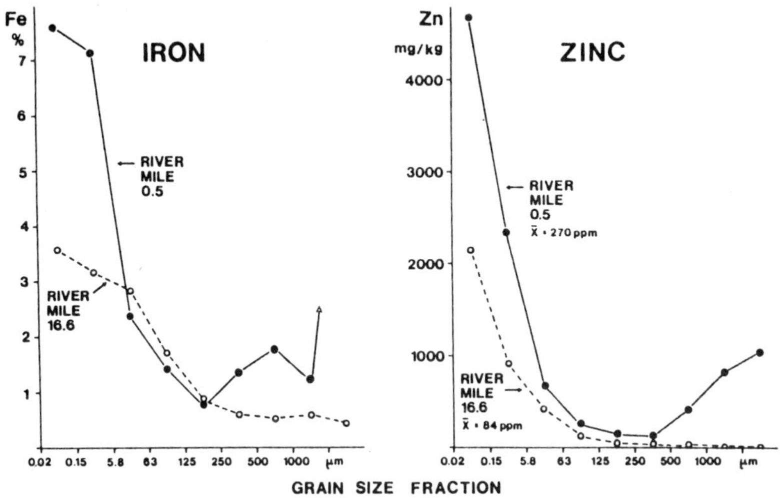

Figure 1 Grain size distribution of iron and zinc in two sediment samples from Saddle River, upstream (mile 16.6) and downstream (mile 0.5) from an urbanized area of Lodi, New Jersey. (Drawn after data of Wilber & Hunter /122/; from Förstner /35/)

would more accurately separate individual particles according to their grain size. However, this is not the case and "coating", for example of iron/manganese oxides, carbonates, and organic substances on relatively inert material in respect to heavy metals, such as quartz grains, act as substrates of heavy metals in coarser grain size fractions. Furthermore, an aggregate formation of finer-grained substances occurs in which the relatively metal-rich clay particles participate as well.

However, it must be stated that the selection of certain grain size fractions - especially in the tracing of sources of pollution - remains controversial for the following reasons (see Figure 1):

(i) Since the larger sediment fractions are less affected by scour and transport, they may best reflect the effect of urbanization on the distribution of heavy metals over an extended period of time at a given location;

(ii) the fine sand fraction (approximately 20μm to 200μm diameter) seems to be of particular interest for the differentiation of natural and pollutant metal transport, firstly because it comprises most of the total sediment, and secondly because there are most obvious differences between individual metal examples (in Figure 1, Fe is not affected in this grain size interval by man-made effects, whereas the typical "anthropogenic" metals, e.g. Pb, Cr, and Zn are significantly enriched in the same grain size fractions of the polluted material).

(iii) The silt and clay fractions comprise the major carriers for both natural and anthropogenic metal compounds (although in relatively small proportions); they are widely distributed with the sedimentation area and are least affected by grain size effects.

3.2. Grain-size corrections

Different methods have been used for grain size corrections (25,41,105); these procedures will reduce (not eliminate) the effect of the fraction of the sediment that is largely chemically inert, including mostly the coarse-grained quartz, feldspar and carbonate minerals, and will increase the substances active in metal enrichment, such as hydrates, sulphides, amorphous and organic materials. In the assessment of metal pollution at a specific locality, the extrapolation techniques both for grain size and specific surface areas of the material require many (10-15) samples. In addition, the calculation of the regression line is a time-consuming and inaccurate procedure. It is often not possible to determine the regression line owing to the limited range in grain size or specific surface area of a sediment at a certain locality. The quartz correction method (114) in some cases has to be extended to dilution effects caused by feldspar, carbonates and opaline silica (106). Normalizing anthropogenic metal concentrations to the contents of "conservative" elements such as Al, K, Fe, Sc and Cs, thus on the percentage of clay minerals, has one major disadvantage in that it gives a ratio value instead of absolute concentration data. But it can be attempted to standardize the contents to a reference material (e.g., average shale) defining the enrichment factor, e.g. of the element "i" in the sample as the concentration ratio of "i" to Al(C_i/C_{Al}) compared with the ratio in the standard material. Due to the close geochemical relation between aluminium and scandium the measurement of the latter component seems to be recommendable for the neutron activation analysis (108,115).

The methods based on the separation of grain size are advantageous because only few samples from a particular locality are needed. However, separation of the fraction less than 2µm in settling tubes (7) is time-consuming (the remobilisation of metals, as has sometimes been suspected, is relatively low;/67/). The fractions less than 173µm and 204µm contain considerable percentages of large grains that are usually low in trace metals. Therefore, the fractions <20µm or <63µm are recommended for the following reasons: (i) trace metals have been found to be present mainly in clay-silt particles; (ii) this fraction is most nearly equivalent to the material carried in suspension (the most important transport mode of sediments by far); (iii) sieving does not alter metal concentrations, particularly when water of the same system is used. While numerous metal studies have already been performed on the <63µm fraction (38,41), it has been effectively argued in favour of the <20µm fraction from estuarine sediments, where a better correlation with "conservative" elements is found than for the fraction <63µm (1).

The Working Group "Aquatic Sediments" of the German Water Chemistry Association has recommended the use of either the separated grain size fraction <20µm (sieving in ultrasonic bath) or the normalization for conservative elements (preferably alumium, scandium or cesium).

3.3. Initial survey of the man-induced metal fraction by solvent leaching

The assessment of both sources and environmental effects of trace metals in particulate matter must predominantly consider the most mobile fraction of these elements, which is that introduced by man's activities and bound to the sediment in sorbed, precipitated or co-precipitated (carbonates and hydrous Fe-Mn oxides) and organically complexed forms. Goldberg (47) suggested that a large portion of the "hydrogenous" metal fraction in recent sediments can be dissolved by specific attack with dilute acids; a residue of relatively inertly bound metals from the natural rock and soil detritus ("residual fraction") can be found predominantly in the coarser grain sizes.

The metal data thus obtained by acid extraction should compare favourably with data from the separation of the fine-grained fraction of sediments. According to a more recent proposal of Bopp & Biggs (11) "environmentally active" trace metals are defined as "those cations that can be separated from the silt and clay fraction of the dried and disaggregated sediment by leaching with 10 percent hydrochloric acid at 70°C for 96 hours".

Studies by Cross et al. (21) on the distribution of Mn, Fe and Zn in sediments of the Newport River Estuary, North Carolina, and by Gross et al. (48) on marine waste deposits of the New York Metropolitan Region were among the first to consider the chemical fractionation of non-detrital/residual bound metals in polluted sediments by extraction with 0.1 M and 1 M hydrochloric acid, respectively. In more recent studies other acids or acid concentrations have come into use for the general differentiation of metal species associated with hydrous oxides, carbonates, sulphides and surficially sorbed materials from the residual fraction, for example, 0.1 M nitric acid (69), 0.3 M hydrochloric acid (85), 0.5 M HCl (2,18), and dilute (25%) acetic acid (82), as well as the chelators EDTA (ethylene-diamine-tetra acetic acid) and DTPA (di-ethylene-triamine-penta acetic acid) commonly used in soil research (45). It seems that these techniques "provide a rapid, inexpensive way of establishing the gross degree to which a sediment population has been subjected to trace metal pollution from the overlying water" (18).

4. METAL SORPTION ON SEDIMENT PARTICLES

Interactions between particulates and metals play an important role in the regulation of dissolved metal concentrations (the most bioavailable). Moreover adsorption is the first step in the ultimate removal of trace metals from the hydrological cycle; the ultimate sink being the oceanic sediments. Changing environmental parameters like pH (acid rain), chlorinity (in estuaries) or the discharge of complexing agents (NTA) all affect the adsorption/desorption processes.

The forms of sorption most commonly encountered in connection with solute transport in soil and sediment are adsorption, chemisorption and ion exchange. There should be no difference between "sorption" and "coprecipitation" (65,76). For example, experiments on cadmium adsorption/coprecipitation with amorphous iron oxyhydrate have clearly shown that the removal efficiency is essentially identical whether the iron is precipitated before the trace metal is added (adsorption) or afterwards (coprecipitation) (9). Physical adsorption on the external surface of a particulate is based on Van der Waals forces of the relatively weak ion-dipole or dipole dipole interactions. Additional reactions can occur with physical sorption on the particle's inner surface or in pores. Chemical adsorption is characterized by the formation of chemical associations between ions or molecules from solution and the surface particles.

Typically adsorption of metals increases from near nil to near 100% as pH increases through a critical range 1-2 units wide (/10/ see Figure 2). This means that a small shift in pH in surface water, as may occur in lakes, causes a sharp increase or decrease in dissolved metal levels. The onset of the adsorption is characteristic for each metal. Under normal conditions in surface water (and at same concentrations of adsorbates and adsorbents) lead and copper are more strongly bound to hydroxidic surfaces than zinc and cadmium. It is important to note, that since the location of the pH-adsorption-"edge" depends on adsorbent concentration (see Figure 2 with different concentrations of amorphous iron oxyhydrate substrate) a statement such as "cadmium adsorbs onto iron oxide at pH 7" is meaningless unless the concentration of adsorbent (and adsorbate) is known (9).

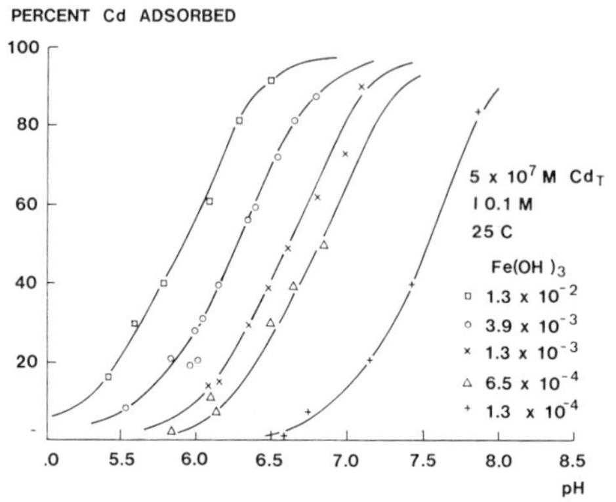

Figure 2

The effect of different adsorbent concentrations - amorph. $Fe(OH)_3$ - on the fractional adsorption (adsorption edges) of cadmium. After Benjamin & Leckie (10).

The solid phases interacting with dissolved constituents in natural waters consist of a variety of components including clay minerals, carbonates, quartz, feldspar and organic solids. These components are usually coated with hydrous manganese and iron oxides, and by organic substances.

Iron and manganese oxides can be present in X-ray amorphous, microcrystalline and in more "aged" crystalline forms both as coatings or as discrete phases. Fe/Mn oxyhydrates are present in all parts of the hydrological cycle. Allochthonous particles in river systems already contain (aged) Fe/Mn oxides, additionally they are formed in stratified lakes, estuaries, and at the sediment/water interface. A further site, particularly for the deposition of Mn-oxide, is the surface of carbonate minerals where there is a microzone of higher pH (77). It is mainly the differentiation under changing redox conditions which determines the relative accumulation of trace metals in Fe/Mn concretions in both marine and freshwaters.

Organic surface for metal sorption could arise in three possible ways (54): (i) from organisms such as bacteria and algae; (ii) by the breakdown of plant and animal material and by the condensation of lower-molecular weight organics (this material would characteristically have a high molecular weight and possess many of the properties of polyelectrolytes and colloidal materials); and (iii) by lower-molecular-weight organic matter being sorbed onto clay or metal oxide substrates (22,23). Although the difference between these three surface types is not well understood with respect to trace metal uptake, there is a general agreement that at least one major binding mechanism involves salicylic entities; other stronger binding entities (such as peptides) may also be present in some systems (55). Data given by Hunter (60) and Tipping (116) suggest that at least part of the organics adsorbed onto the particulate matter in natural waters has carboxylic and phenolic functional groups available for binding with trace metals. The bulk of soluble organic matter has characteristics similar to those of fulvic acid (102). The trace metal adsorption capacity of organic matter is generally between that for metal oxides and clays (50).

The sorption of metals on sediments is complicated because the various organic and inorganic ligands in solution and the various components in the sediments all compete for complexation of the trace metals. With regard to competing ligands in solution, two situations can be envisaged as follows (24):

(a) The complexing ligand (for example, chloride) is not adsorbed onto the sedimentary components, and thus it competes with surfaces for coordination of the metal ion. Increasing concentration of the ligand causes a decrease in adsorption as exemplified by the chlorinity increase in estuaries (Figure 3). It should be noted, however, that the concentration of the suspended matter may also increase (turbidity maximum, erosion of sediments) causing an increase in adsorption despite a rise in chlorinity (103).

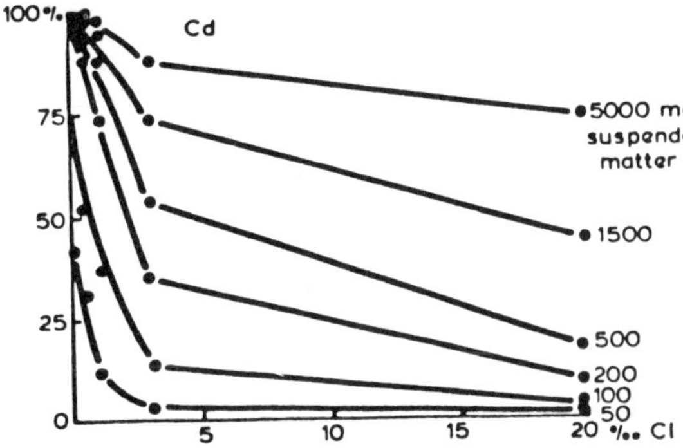

Figure 3

Adsorption of cadmium (5 μg/l) on sediment from the Rhine River: influence of chlorinity and suspended matter concentration (from Salomons /103/)

(b) The complexing ligand is adsorbed onto surfaces. If this is the case, the adsorption of metal ions can be significantly enhanced. This appears to be the case for EDTA (24) and NTA (59a). In this way, the speciation of trace metals in solution determines the particulate metal speciation.

Investigations on the adsorption kinetics and reversibility of adsorption equilibria (81) indicate, that reversibility of the sorption processes is strongly different in natural mixtures of solid substrates from the experience on well-characterized solid surfaces. Figure 4 gives the results of two seperate experiments with lead and cadmium on sediment samples from the oxidized surface layer of mudflats in the South San Francisco Bay estuary (81). Reaction systems were equilibrated for 24 h at the appropriate pH for approximately 90% metal adsorption as determined by prior experiments; the pH was then lowered to that approximately 20% to measure reversibility of sorption reaction. The results indicate slow release of adsorbed cadmium within a time frame of 96 h, whereas lead was substantially non-labile over the 264-h duration of the experiment.

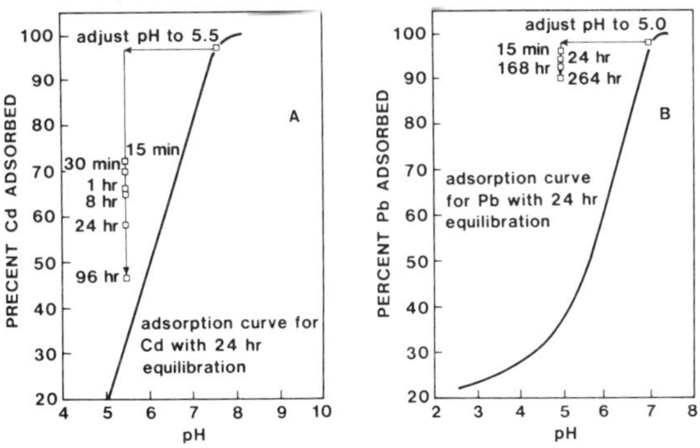

Figure 4

Kinetics of desorption from unaltered salt marsh sediment. A Cadmium B Lead

(from Lion et al. /81/).

While the reduced reversibility is mainly the result of the effects of hydrolysis and specific adsorption there are other phenomena which characteristically affect the kinetics of metal release from particulate matter:

The typical effects of the earliest stages of "diagenesis" (involving transformations of the organic matter,"aging" of mineral components and formation of new equilibria between solid and dissolved species) have been demonstrated by Salomons (103) with respect to the behaviour of trace metals at the freshwater/seawater interface (see also Figure 3 for the antagonistic effects of chlorinity and suspended matter concentration). Desorption was studied by adding cadmium and zinc to suspended matter in riverwater (Table 1). After adsorbing periods of 1, 3, 8, 24, and 60 days, NaCl was added to the suspension to increase the chloride concentration to 1.9% (approx. seawater composition). After an adsorption period of only one day, 24% of the adsorbed cadmium and 60% of the adsorbed zinc remains bound to the sediment; after 60 days 40% of the cadmium and 88% of the zinc bound to the sediment is not released after NaCl treatment.

Table 1: Percentage of cadmium and zinc not released from river suspended matter after treatment with NaCl solutions. From Salomons (103)

Adsorption time	1 day	3 days	8 days	24 days	60 days
Cadmium	24%	30%	33%	37%	40%
Zinc	60%	67%	70%	74%	88%

An extrapolation to geologic time scales can be made by a comparison of the bonding intensity of stable metal isotopes and their instable counterparts - the latter supplied from radioactive emissions of nuclear power and processing plants. Figure 5 shows the effects of sequential leaching of a sediment sample from the lower Rhône River in France. It is demonstrated that the reducing agents hydroxylamine (pH 2) and oxalate buffer (pH 3) extract only 15% of the natural stable manganese while the artificial isotope Mn-54 is mobilized at more than 80% by these treatments. One of the implications of these data is that under reducing conditions there should be a more intensive release of the radioactive manganese nuclide compared to the stable counterparts; also the "availability" of the "younger" Mn-nuclide should be much high than for the stable "geochemical" Mn-isotopes (39).

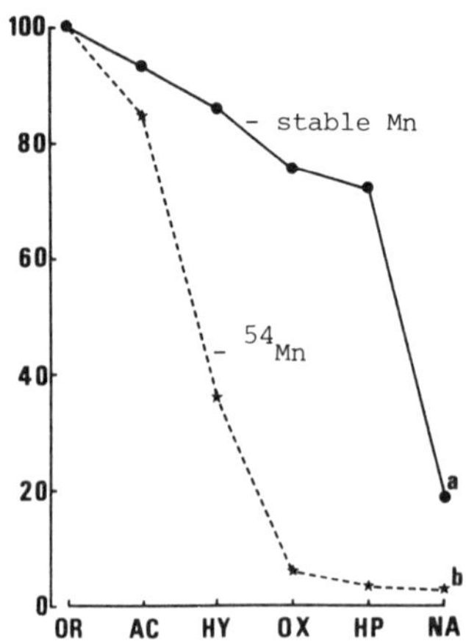

Figure 5

Comparison of chemical extractability of artificial and stable isotopes of Mn from a sediment sample of the Rhône River (39)

OR = original sample (gamma activity=100%)
AC = after treatment with 1 M NH_4OAc pH 7
HY = treatment with hydroxylamine·HCl
OX = treatment with oxalate buffer pH 3
HP = hydrogen peroxide pH 2 treatment
NA = conc. nitric acid

(Details see section 5)

5. CHEMICAL EXTRACTION METHODS FOR METAL SPECIATION IN SEDIMENTS

There are five major mechanisms of metal accumulation in sedimentary particles (46):

- adsorptive bonding on fine-grained substances
- precipitation of discrete metal compounds
- co-precipitation of metals by hydrous Fe and Mn oxides
- associations with organic molecules
- metals incorporated in crystalline minerals

Table 2 gives a compilation of the various mechanisms, products, and substrates of metal enrichment and their estimated significance in aquatic sediments; this scheme includes all major types of metal associations occurring in both natural and man-affected water systems:

Table 2: Mechanisms and substrates of metal enrichment in aquatic sediments.

	Rock debris, solid waste material, incl.organics	Metal -hydroxide -carbonate -sulfide	Organic substance (reactive)	Hydrous Fe/Mn-oxide (mainly authigenic)	Calcium carbonate
Inert bonding	xxx			x	x
Precipitation		xxx		xx	xx
Adsorption (phys.)	x	x	x	x	x
Coprecipitation		xx		xxx	xx
Coating (surface)			xxx	xxx	x
Flocculation			x	xx	

It is suggested, that the inert bonding of metals, e.g. in clay minerals and heavy minerals is the dominant effect of metal accumulation in natural sediments. In man-affected systems, enrichment processes mainly occur via precipitation of discrete metal compounds, taking place when external factors change; these precipitation barriers commonly arise because of changes in pH, oxidation potential, or in concentrations of precipitating substances (101). With respect to surface-associated mechanisms, the accumulation on organic substances and hydrous Fe/Mn oxide seems to be particularly important. Flocculation and coagulation processes at the river/sea interface mainly affect removal of iron as colloids (110), whereas dissolved organic matter behaves widely conservate during estuarine mixing (90), because the interfacially-active humic acids that coagulate in association with iron colloids comprise only a minor fraction of the dissolved organic matter (61).

There are still controversial data on the effect of organic matter for metal binding in natural sediments. Sorption experiments on river sediments, which had been treated with hydrogen peroxide (to partially eliminate organic matter) showed a significant decrease of the sorption capacity for copper, zinc, cadmium and lead (112). Studies on the bonding of radionuclides on different organic substrates indicate both an increase and decrease of the sorption characteristics - depending from ligand concentrations and metals involved (19). In the limnic environment depositon of organic substances on solid surfaces usually coincides with a decrease of the sorption of heavy metals on these surfaces (6).

5.1. Single leaching steps

A compilation of single leaching steps involved in common extraction procedures is given in Table 3 and in Figure 6 (95,104):

In determining the amount of exchangeable metal ions, an extractant is used which contains cations more strongly bound to the exchange positions as compared with the metal; common extractants in use are $BaCl_2$, $MgCl_2$ and NH_4OAc. The ion-exchangeable fraction of trace metals is still ill-defined as is indicated by significant metal releases in replicate experiments (104).

To release metal ions weakly bound to specific inorganic sites (e.g., carbonates) extraction with 0.4 M HOAc, HOAc/NaOAc (pH 5), or complexing agents of moderate ligand strength has been recommended (95). A single extraction method for the determination of trace metals associated with carbonates makes use of an acidic ion-exchanger (26).

Dissolution of the amorphous (or "poorly ordered") inorganic phases can be promoted by reducing the iron and manganese to lower valency states, e.g. through reactions with hydroxylamine, oxalate or dithionite ions. Oxalate treatment did not appear to attack crystalline material, but it has been shown that the degree of interaction varies with illumination (109, 95). For routine investigations an extraction with 0.3 M HCl can be recommended because this procedure requires fewer manipulations (45).

The metal fraction associated with organic matter has been extracted through oxidation of this component with NaOCl or H_2O_2, by dissolving the fraction using $Na_4P_2O_7$, through proton displacement (with 0.5 M HCl) or by chemical competition (e.g., using 0.05 N EDTA); the apparent ability of both EDTA and HCl to release metal ions associated with both organic and oxide phases has led to these being used as extractants for the evaluation of the total nondetrital fraction (95; see also section 3.3.).

For the dissolution of detrital or lattice-bound components a combination of acids has generally been used. Nitric acid promotes destruction of organic matter, but when used alone or in conjunction with HCl or $HClO_4$, the procedure does not completely dissolve some types of silicate minerals.

The diagram of Pickering (95) in figure 6 implies - dashed segments indicate areas of uncertainty - that extraction procedures may only crudely differentiate between the different forms of selected species in sedimentary material. In particular, despite of clear advantages of a differentiated analysis over investigations of total sediment, it should be clearly pointed out that the various extraction steps are not as selective as sometimes stated. Several problems are chosen for illustration of this point (37,104):

1. Labile metal phases could be transformed during sample preparation, which can occur especially for samples from reducing environments (67).

2. Processes of readsorption and precipitation have to be considered, particularly during extraction with ammonium acetate and hydrogen peroxide (98).

3. Reactions are influenced by the duration of the experiment and by the ratio of solid matter to volume of extractant. Some of the extractants are more specific than others, such as shown by repeated experiments (56). A too high solid content, together with an increased buffer capacity may cause the system to overload; such an effect is reflected, for example, by changes of pH-values in time-dependent tests (94).

4. There are major objections to the use of bariumchloride/triethanolamine (/64/; the latter compound is used to raise the pH to 8.1) or magnesiumchloride (46) for the determination of exchangeable trace metals; there are chelating effects of the organic agent and the formation of soluble metal-chloro-complexes. The neutral ammonium-acetate solution (1 molar) seems to offer an acceptable compromise for the estimation of ion-exchange-

Table 3: Methods for the extraction of metals from major chemical phases in sediments (examples). After Salomons & Förstner (104).

Method	Reference
Adsorption and cation exchange	
$BaCl_2$-triethanoleamine pH 8.1	Jackson (64)
$MgCl_2$	Gibbs (46)
ammonium acetate pH 7	Engler et al. (32)
Carbonate phases	
CO_2-treatment of suspension	Patchineelam (93)
acidic cation exchanger	Deurer et al. (26)
NaOAc/HOAc-buffer (pH 5)	Tessier et al. (113)
Reducible phases (in approximate order of release of iron)	
acidified hydroxylamine (+ 0.01 M HNO_3)	Chao (16)
amonium oxalate buffer	Schwertmann (109)
hydroxylamine-acetic acid	Chester & Hughes (17)
dithionite-citrate buffer	Holmgren (59)
Organic fraction (incl. sulfides)	
H_2O_2-NH_4OAc (pH 2.5)	Engler et al. (32)
Organic solvents	Cooper & Harris (20)
0.1 M $NaOH/H_2SO_4$	Volkov & Fomina (120)

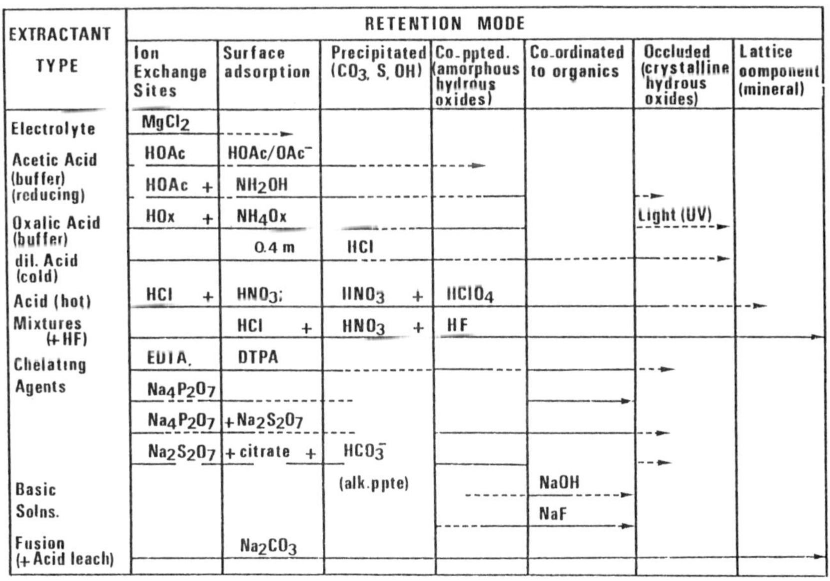

Figure 6: Schematic representation of the ability of different extractant solutions to release metal ions retained in different modes or associated with specific soil and sediment fractions. Dashed segments indicate areas of uncertainty. From Pickering (95), CRC Crit.Rev.Anal.Chem.,Nov. 1981

able trace metal contents (104).

5. Basic metal oxides may be formed during initial high pH conditions, for example, for humic extraction with sodium hydroxide or sodium pyrophosphate (12).

6. Difficulties occur with all carbonate extractions: reprecipitation under oxygenated conditions following CO_2 treatment, adsorption of organic compounds on exchanger resin, pH effects on labile oxyhydrates. However, the major advantage of this step lies in the reduction of the buffer capacity of the sample before applying the following leaching steps (104).

7. Problems in the acid reducing steps are encountered particularly with the utilization of the dithionite/citrate extraction (113): contamination by metals in the reagent, clogging of the burner during atomic absorption analysis, decomposition of the extractant and formation of sulfides (most of them being highly insoluble).

8. Treatment with hydrogen peroxide affects both labile as well as relatively stable bound metal components (31).

9. Treatment with organic solvents is poorly suited for routine application (104).

10. The differentiation of organic and sulfidic metal associations, e.g. in the peroxide fraction, is not as yet possible (13).

5.2. Sequential extraction procedures

In connection with the problems arising from the disposal of contaminated dredge material, sequential extraction procedures have been developed, which include the successive leaching of metals from interstitial waters, and from ion exchangeable, easily reducible, organic and residual sediment fractions (32,45,49). Based on our experience and an evaluation of the relevant literature, we have suggested a standard extraction procedure consisting of three steps for the differentiations of the three major groups of phases in sediments (104)

- extraction with acidified hydroxylamine hydrochloride at pH 2 (including the extraction of exchangeable cations and of carbonate-bound metals);
- extraction with acidified hydrogen peroxide, followed by an extraction with ammonium acetate to remove any readsorbed metal ions;
- total analysis ($HF/HClO_4$) of the residue remaining after the first two extractions.

This scheme, including an additional step for exchangeable cations was used for the study of sediments from 18 different river systems (104). In figure 7 the rivers are arranged according to their approximate geographical position from North to South. (Most tropical rivers contain low amounts of cadmium and here it was not possible to obtain a reliable speciation; for the Rio Magdalena and the Orinoco River, insufficient material was available for the determination of the exchangeable fraction; the acidified hydroxylamine step therefore contains also the "exchangeable metals"). The results show an increase in the relative amount of metals present in the resistant (lithogenous) fraction of the less polluted river systems. In nearly all examples the "reducible phase" is an important sink for trace metals; copper more than other trace metals shows a tendency to refer the "organic phase".

In extension of this sequence and using the experience of the work from Tessier et al. (113) a six-stage chemical leaching process was developed for the assessment of the environmental compatibility of dredged materials that avoids some of the inadequacies of earlier experiments (37,43):

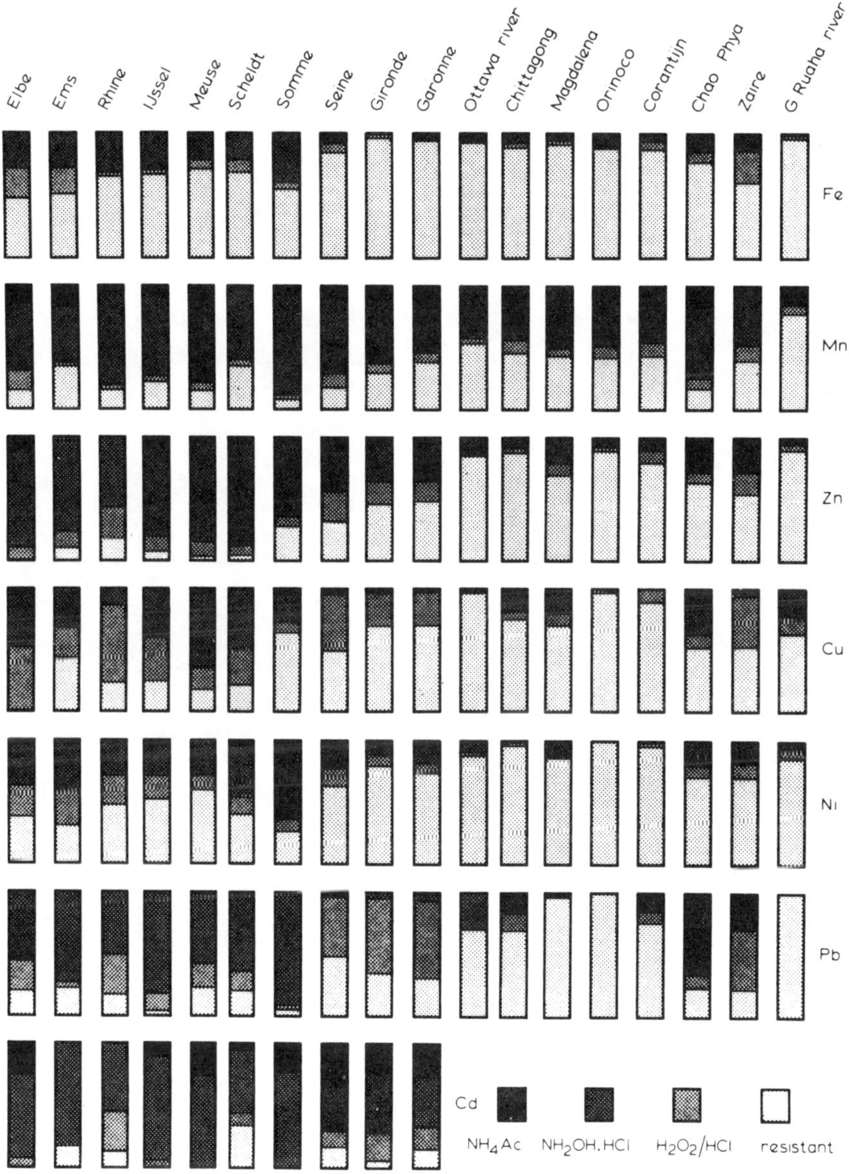

Figure 7

The speciation of trace metals in 18 different river sediments according to the "standard extraction method" (after Salomons & Förstner /104/).

Fraction 1: Exchangeable cations
The sediment sample was extracted for 2 hrs shaking time with 1 M ammonium acetate pH 7, solid/solution ratio 1:20 (Tessier et al. /11/: 1 N $MgCl_2$, pH 7, 10 min)

Fraction 2: Metals bound to carbonate
The residue from fraction 1 was leached for 5 h with 1 M sodium acetate adjusted to pH 5.0 with acetic acid.

Fraction 3a: Easily reducible phases (Mn oxide, partly amorphous Fe-oxide)
The residue from fraction 2 was leached for 12 hrs with 0.1 M hydroxylamine hydrochloride + 0.01 M nitric acid, pH 2, dilution 1:100.

Fraction 3b: Moderately reducible phases (amorphous + poorly cryst.Fe-oxide)
The residue from fraction 3a was extracted for 24 hrs with 0.2 M ammonium oxalate + 0.2 M oxalic acid, pH 3, dilution 1:100
(Fraction 3 after Tessier et al. /11/: The residue from fraction 2 was extracted for 6 hrs at 96°C with 0.04 M $NH_2OH \cdot HCl$ in 25% (v/v) HOAc)

Fraction 4: Metal bound to organic matter and sulfides
The residue from fraction 3 was extracted at 85°C for 5 hrs with 30% H_2O_2, adjusted to pH 2 with HNO_3 and then at room temperature with 3.2 M ammonium acetate in 20% nitric acid.

Fraction 5: Residual metals (e.g. detrital silicates, crystalline Fe-oxide)
The residue from fraction 4 was digested with a 5:1 mixture of hydrofluoric and perchloric acids (Tessier et al. /11/) or with conc. nitric acid, 120°C, dilution 1:100 (37).

Examples for the solid phase partition of Cd have been given in Table 4 for samples from areas of major dredging activities. Compared to other metals, such as Pb and Cu, cadmium is characteristically enriched in the more mobile fractions "cation exchange" and "easily reducible phases" and is, therefore, more mobile than most of the other heavy metals. The data from Table 3 also suggest a general decrease in the residual bonding form, that is, the predominantly inertly fixed metal content, as the anthropogenic metal enrichment increases. At least in this general form it may be estimated that the remobilization of Cd under changing environmental conditions and potential uptake by biota should also increase in the same direction.

Table 4 Chemical forms of cadmium in sediments and suspended matter (42).

Sediment fraction$^{(a-e)}$	Weser (1)	Estuary (2)	Rhine R. (3)	Rotterdam Hr. (4)	Neckar R. (5)
Cation exchangeable a)	8%	18%	27%	12%	13%
Easily reducible b)	32%	49%	42%	61%	64%
Moderately reducible c)	45%	14%	2%	13%	10%
Organic/sulfidic d)	12%	17%	17%	12%	12%
Residual fraction e)	3%	2%	12%	2%	1%
Total Cd (mg/kg)	3.0	4.3	9.0	18.1	33.6

(1) Dredged sediment; (2) suspended matter; (3) station Wesel (Dutch/ German border); (4) Broekpolder; (5) lock reservoir Lauffen

This scheme has been applied, apart from the major studies on river sediments and dredged materials, to demonstrate the relative mobility of particle-associated trace metals in lake sediments affected by acid rain (99,100), in sand filter column experiments (43), and - with modifications - for the speciation of reducible metal compounds in pelagic sediments (40,94).

Respective data on solid speciation of Cd, Zn, and Pb are given in Table 5 for matrices such as street dust (A) and roadside soil (B), fresh (C) and incinerated sewage sludge (D):

Table 5 - Chemical extraction of cadmium, zinc and lead from different fractions of solid waste materials. A = street dust; B = roadside soil (53); C = fresh sewage sludge (42); D = incinerated sewage sludge ash (44).

	cadmium				zinc				lead			
	A	B	C	D	A	B	C	D	A	B	C	D
Exchangeable	20%	26%	17%	-	2%	3%	4%	-	2%	1%	-	-
Carbonate fr.	38%	24%	61%	-	44%	31%	69%	-	43%	26%	11%	1%
Reducible fr.	28%	25%	6%	1%	43%	34%	14%	4%	38%	44%	29%	1%
Org./sulfide	8%	8%	11%	10%	7%	9%	7%	6%	7%	12%	16%	1%
Residual fr.	6%	18%	5%	89%	4%	23%	6%	90%	10%	17%	44%	98%
µg/g dry mat.	4.0	2.2	1.8	10.9	465	345	1267	1.34%	1500	600	110	2275

A study made by Harrison et al. (53) with 6 street dust (A in Table 5) and 4 roadside soil (B in Table 5) samples indicates that cadmium is there to an appreciable proportion in the exchangeable fraction while zinc and lead is predominantly associated with carbonate and Fe-Mn oxides (copper is largely in organic associations). The high cation and chloride concentration (1 M $MgCl_2$) in the exchange solution may reflect conditions in soils contaminated with de-icing salt (53). In comparison to the street dust leaching data the experiments on roadside soils indicate significantly lower extraction rates in the more labile phases and an increase of the residual metal percentages. This is connected with the lower total metal contents in the roadside soil samples compared to the dust specimen. Despite the relatively low concentration of cadmium in the sewage sludge sample (C; sample from a treatment plant at Landau/Pfalz /42/) there is a significant shift to higher percentages in the carbonate fraction; this is also valid for zinc whereas lead is enriched in the organic and residual forms, compared to the atmospherically induced samples (A and B). The general experience that the more enriched elements are also the more reactive ones (see above) does not fit to all waste materials. This is exemplified by the data from incinerated sludge ash (sample D; Hamilton Municipal Incinerator /44/), where the elements are highly enriched but are rather immobile with potentially toxic metals Cd, Zn, and Pb ca. 90% or more in "unavailable" forms (44).

With regard to the various influencing factors affecting the chemical associations of trace elements in less contaminated as well as in polluted sediments, investigations on sedimentary core profiles seem to be particularly useful. Apart from the characteristic developments in the chronological sequences, there is a major advantage in comparing samples when they have been treated according to the same extraction procedures and analytic methods. This also pertains to filter column experiments, as long as the filling material can be considered homogeneous with respect to grain size and chemical composition (43).

5.3. Specificity of extraction procedures

Although "selectivity" for a specific phase in a strictly thermodynamic sense cannot be expected for these procedures, there are differences in the "specificity" of the various extractants used in the sequential extraction procedures. We have tested the sequential procedure of Tessier et al. (113 using well defined mineral phases such as Ca, Mn, Cd and Pb carbonates, Fe and Pb sulfides, Fe oxides (goethite and hematite), Fe phosphate (vivianite), chloritic clay minerals as well as a relatively simply mixed oxidic phase: deep-sea Fe/Mn nodule material. After each extraction step the residual solid phase was analysed by scanning electron microscopy with energy dispersive X-ray equipment (EDAX).

It is shown in Figure 8 (97) that the carbonate phase is undoubtedly attacked during step 2 of the extraction procedure; however , step 1 (ammonium acetate pH=7) also partially dissolves this phase (as has already been mentioned by Jackson /64/). In the case of deep-sea Fe/Mn nodules step 3 was extremely selective: all the manganese and 90% of the iron was extracted; the residual iron is probably due to various impurities such as clay minerals. Cadmium and part of the zinc was found in the exchangeable phases (steps 1 and 2) and are probably adsorbed on the hydroxy-oxides of Fe and Mn, since the latter are known for their high adsorption capacity. An artificial amoprphous FeS sample was sequentially leached: as one would expect 60% of the iron was solubilized by step 4, however 20% was released by step 2. Similarly, using a sample of galena (PbS) 80% of the lead is extracted by step 4 as well as the Cd, Zn and Ni; the remaining 20% of lead and all of the iron and manganese is extracted by step 2 and 3, probably due to an oxidic impurity coating the galena particles. For crystallized iron oxides such as goethite or hematite only a slight attack using step 3 was observed, while iron is principally extracted only during step 5 (acid digestion); significant proportions of cadmium and zinc were observed in the carbonate fraction. Chlorite - an example for the clay mineral group - is attacked by the last step; however, step 3 (hydroxylamine hydrochloride) released about 15% of the iron; this has been proposed already by Chester & Hughes (17). Part of the Cd, Pb and Cu is leached during steps 1 and 2, indicating that these metals are adsorbed onto the particulate phases. Iron phosphates such as vivianite are attacked during step 3 to 5 to an equal extent; there is, however, a significant portion of cadmium, zinc and nickel being released already in steps 1 and 2 (exchangeable and carbonate fraction).

Recent investigations by Kersten (74) on dredged harbour sediments have clearly shown, that the sample preparation is distinctly affecting the speciation of (at least) iron and manganese in anoxic material. Exchangeable manganese (step 1) was 10-20% higher in fresh samples than after freeze-drying; in fresh samples approximately 5% of the iron was associated with organic and sulfidic phases (step 4) whereas after freeze-drying no more Fe was found in this fraction. Obviously these effects are not valid for older reducing sediments, where more stable iron sulfides have been formed diagenetically. Initial data of Kersten (74) from extraction experiments on anoxic sediment under oxygen-poor conditions (N_2-atmosphere in glove-box) indicate, that the proportion of iron in the carbonate phase was approximately 30% compared to 13-18% after freeze-drying of the samples. Oxidation under normal preparation (without oxygen-exclusion) shifted the carbonate-bound iron fraction to the moderately reducible phases (extracted by treatment with oxalate buffer solution). These data demonstrate - as has already been suggested by Filipek (33) - that anoxic sediments must be extracted with great care under oxygen-free conditions at least for the first two steps, and the respective leaching agents have to be de-aerated prior to the contact with the original, wet samples.

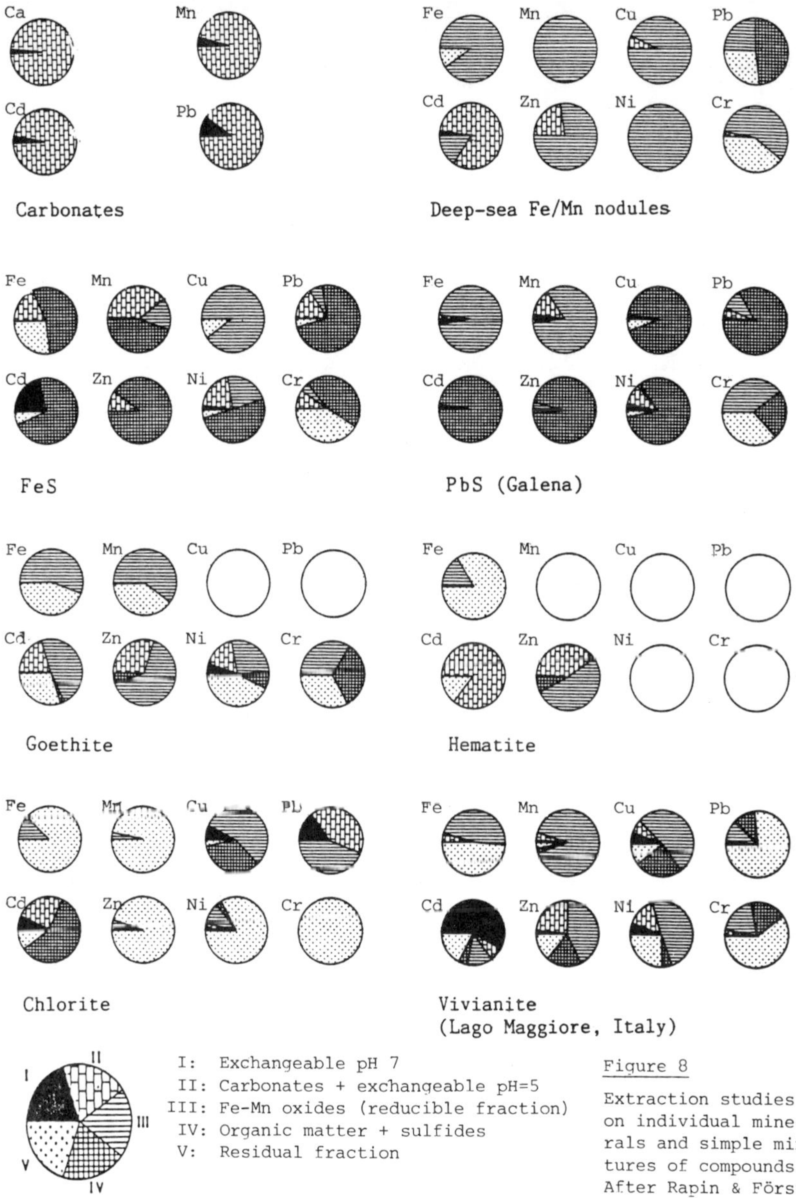

Figure 8

Extraction studies on individual minerals and simple mixtures of compounds. After Rapin & Förstner (97)

6. ESTIMATION OF MOBILITY AND BIOAVAILABILITY FROM SOLID SPECIATION DATA

The mobility of elements in sediments is affected by pH, redox potential, and complexing ligands (both organic and inorganic). It has been stressed by Plant & Raiswell (96) that the predominance of simple mineral-solution equilibria is reasonable for the major elements in the surface environment, while the behaviour of many trace elements is more complex, and is also determined by coprecipitation and surface effects and interactions with organic phases. However, typical developments can be estimated from the simple scheme in Table 6 (partly after /96/), including data from soil studies (86,68) and recent estimations on the effects of acid precipitation (14):

Table 6: Relative mobilities of elements in soils and sediments

Relative mobility	Electron activity		Proton activity	
	reducing	oxidizing	neutral-alkaline	acid
very low mobility	Al, Cr Mo, V, U, Se S, B Hg, Cu, Cd, Pb	Al, Cr Fe, Mn	Al, Cr Hg, Cu Ni, Co	Si
low mobility	Si, K, P Ni, Zn, Co, Fe	Si, K, P Pb	Si, K, P Pb, Fe, Zn, Cd	K Fe(III)
medium mobility	Mn	Co, Ni Hg, Cu, Zn, Cd	Mn	Al Pb, Cu, Cr, V
high mobility	Ca, Na, Mg, Sr	Ca, Na, Mg, Sr Mo, V, U, Se	Ca, Na, Mg, Sr	Ca, Na, Mg, Sr Zn, Cd, Hg Ni, Co, (Mn)
very high mobility	Cl, I, Br	Cl, I, Br S, B	Cl, I, Br S, B Mo, V, U, Se	Cl, I, Br S, B

It can be expected that changes from reducing to oxidizing conditions will strongly increase the mobility of metals such as Mo, V, U, and Se, significantly of Hg, Cu and Cd, and probably to some extent also of Zn, Co, Ni, and Pb. On the other hand, mobility is characteristically lowered for Mn and Fe under oxidizing conditions. Geochemical mobility in response to environmental acidification will significantly increase, if the following metals are present in the non-crystalline fraction of the sediment (14): Al, Mn, Zn, and perhaps Cd, Co, and Ni; on the other hand the elements exhibiting anionic species, such as S, As, Se, Cr, and Mo are appreciably solubilized, e.g. from fly ash sluicing/ponding systems, at neutral to alkaline pH-conditions (29,11).

From experiments described in the literature Jonasson (68) established a probable order of binding strength for a number of metal ions onto humic or fulvic acids: $Hg^{2+} > Cu^{2+} > Pb^{2+} > Zn^{2+} > Ni^{2+} > Co^{2+}$. It has been suggested that natural dissolved organic matter, such as humic material or amino acids present usually in the sea but also in certain inland waters, is insufficient to make any significant contribution to the speciation of dissolved Cd, Pb or Zn (92); only the distribution of Cu species seems to be affected markedly by organic complex formation with natural chelators. There are only few data on the effects of higher concentrations of organic ligands in pore waters; here too, a sequence of stability can be established in the order

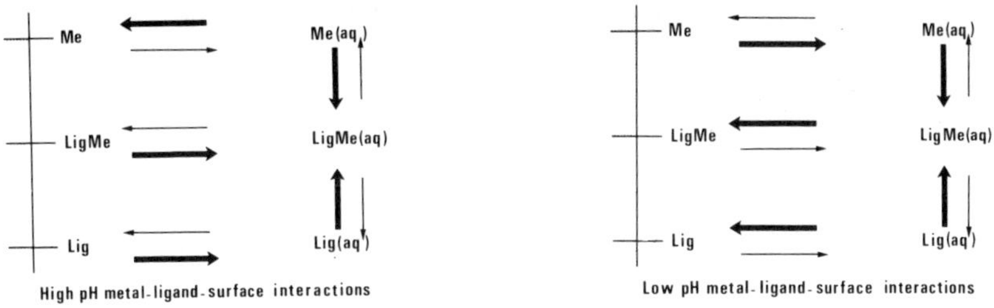

Figure 9: Adsorption of metals onto solid surfaces in the presence of complexing ligands at high pH (left) and low pH (right) conditions (after Benjamin & Leckie/10/).

Cu > Pb > Cd (8). Particularly with respect to stronger complexes, pH-dependencies of the dissolved/solid interactions are quite significant. Schematic representation of these influences in Figure 9 (Benjamin & Leckie /10/) suggest, that enhanced adsorption of a metal can be expected at lower pH-values in the presence of a stronger complex-forming ligand (such as NTA, which is used in some countries already in detergents to substitute polyphosphate).

6.1. Estimation of potential remobilization of metals from sediments

Species differentiations can be used for the estimation on the remobilization of metals under changing environmental conditions:

In the estuarine environment the "exchangeable fraction" might be affected in particular; however, changes of pH and redox potential could also influence other easily extractable phases, e.g. carbonate and manganese oxides. The high proportion of Cd in exchangeable fractions of polluted sediments is reflected in the typical mobilization effect of this element in the estuarine environment (2,107).

Lowering of pH will affect, according to its strength, the "exchangeable", then the "easily reducible" and in case parts of the "moderately reducible" fraction, the latter consisting of Fe-oxyhydrates in less crystallized forms.

Redox changes concern under "postoxic" conditions, during reduction of, for example, nitrate and manganese oxide, the "easily reducible fraction", i.e., metals associated with manganese oxides and partly with amorphous Fe-oxyhydrates; in strongly reducing environments, e.g., in highly polluted sediments, the "moderately reducible fraction" is affected too, especially when the iron is present in the form of coatings (65).

The effects on organically bound metals are more complex; however, it has been argued that this fraction is highly susceptible to environmental changes, especially during early diagenetic reactions, where recycling of mineralized organic matter and pore-fluid transfer processes are controlling the dynamics of pollutants and nutrient in sediments (78).

6.2. Estimation of bioavailability of sediment-associated metals

The availability of trace elements to organisms may be influenced by factors such as (i) the physiological and ecological characteristics of the organisms, (ii) the form of dissolved trace elements, as well as the chemical and physical characteristics of the water, and (iii) the form of trace elements in solids (66). With respect to the solid speciation much work has been done on the relation between chemical leachability and plant uptake of metals. Highly significant dependencies have been found for extractants such as 0.05 M DTPA, EDTA, acetic acid, ammonium acetate and nitrate (70,111,121). It must be mentioned, however, that "the leachable fraction does not necessarily correspond to the amount available to vegetation" (95). One major reason for this is the lack of information about the specific mechanism with which the organisms actively participate in the removing of nutrient from the solid substrate; in the case of plant root activities these interactions with soil components include redox changes (15), pH alterations (89) and organic complexing processes (79).

With respect to aquatic organisms the data of Luoma & Bryan (83) from sediments in 20 estuaries in southern and western England indicate that the concentration of lead in the deposit-feeding bivalve Scrobicularia plana is largely controlled by the concentration in the sediment as modified by the inhibitory effect of iron; the Pb/Fe ratio in 1 N hydrochloric extracts appears to reflect the availability of lead to Scrobicularia. More detailed investigations including statistical analysis of data performed by Luoma & Bryan (84) suggest that the correlation between metal uptake and solid species distribution reflects competition for various sorption sites within the sediment. It is assumed that the availability of zinc to Macoma in San Francisco Bay is controlled by the competitive partitioning of Zn between two forms of hydrous oxides: hydroxylamine hydrochloride-soluble iron oxide and ammonium-acetate-soluble manganese oxide; organic materials affect the bioavailability of Zn when present in concentrations greater than 1 percent.

The present state of knowledge on solid matter speciation of trace metals is still somewhat unsatisfactory with respect to the assessment of the quantitative extent of bioavailable element concentrations. This is mainly due to the effect of competition between adsorption sites on the solid substrate and selective mechanisms of metal uptake by the different organisms - a situation which cannot be improved by more sophisticated extraction data. On the other hand, the usefulness of a differentiated approach on the interactive processes between water/biota and - even only operationally defined - solid phases has been clearly evidenced. The possible environmental implications, e.g., during dredging operations, after land disposal of waste material, from acid precipitation, for redox changes in the subsoil, and organic complexing near sewage outfalls, can be qualitatively estimated, particularly when the hydrochemical conditions are known. With respect to the more important environmental parameters, such as pH and redox potential (105) a weak acid-reducing extractant offers a relatively good indication of the mobile and potentially bioavailable fraction of transition metals, equivalent to the nonapatite inorganic P-fraction of the solid phosphorous compounds (104). Some of the still obvious inadequacies in the correlation of data between the different solid substrates and the organisms (or parts of them) can possibly be reduced by the study of the intermediate water phase - interstitial water - and it seems to be particularly promising to consider the different forms and availabilities of metals in this medium.

Acknowledgements. Dr. W. Salomons, Delft Hydraulics Laboratory (Haren Branch) kindly provided material for the present review. Discussions on the possibilities and limitations of sequential extraction techniques with colleagues from our department in Hamburg and from other institutions were greatly appreciated. Dipl.Ing. M. Kersten provided data on the effect of oxygenation and freeze-drying on labile metal phases in anoxic sediments; analytical work has partly been performed by Ms. I. Brauer from our laboratory.

REFERENCES

1. ACKERMANN, F., BERGMANN, H. and SCHLEICHERT, U. (1983). Monitoring of heavy metals in coastal and estuarine sediments. A question of grain size: <20 µm versus <60 µm. Environ. Technol. Letts. 4: 317-328.

2. AGEMIAN, H. and CHAU, A.S.Y. (1976). Evaluation of extraction techniques for the determination of metals in aquatic sediments. The Analyst 101: 761-767.

3. AHLF, W. (1983). The River Elbe: Behaviour of Cd and Zn during estuarine mixing. Environ. Technol. Letts. 4: 405-410.

4. American Society Civil Engineering (1972). Manuals and Reports on Engineering Practice: Ground Water Basin Management, New York.

5. American Society for Testing and Materials (ASTM) (1979). Annual Book of ASTM Standards, Part 31 "Water", Philadelphia Pa.

6. BACCINI, P., GRIEDER, E., STIERLI, R. and GOLDBERG, S. (1982). The influence of natural organic matter on the adsorption properties of mineral particles in lake water. Schweiz. Z. Hydrol. 44: 99-116.

7. BANAT, K., FÖRSTNER, U. and MÜLLER, G. (1972). Schwermetalle in Sedimenten von Donau, Rhein, Ems, Weser und Elbe im Bereich der Bundesrepublik Deutschland. Naturwissenschaften 59: 525-528.

8. BATLEY, G.E. and GILES, M.S. (1980). A solvent displacement technique for the separation of the sediment interstitial waters. In: R.A. BAKER (ed.) Contaminants and Sediments. Ann Arbor Sci. Publ. Vol. 2: 101-117.

9. BENJAMIN, M.M. and LECKIE, J.O. (1981). Multiple-site adsorption of Cd, Cu, Zn and Pb on amorphous iron oxyhydroxide. J. Colloid Interface Sci. 79: 209-211.

10. BENJAMIN, M.M., HAYES, M.M. and LECKIE, J.O. (1982). Removal of toxic metals from power-generation waste streams by adsorption and co-precipitation. J. Water Pollut. Control Fed. 54: 1472-1481.

11. BOPP, F. and BIGGS, R.B. (1981). Metals in estuarine sediments: Factor analysis and its environmental significance. Science 214. 441-443.

12. DURTON, J.D. (1978). The models of association of trace metals with certain compounds in the sedimentary cycle. In: E.D. GOLDBERG (ed.) Biogeochemistry of Estuarine Sediments. Unesco Paris, pp. 33-41.

13. CALVERT, S.E. and BATCHELOR, C.E. (1978). Major and minor element geochemistry of sediments from Hole 379A, Leg 42B, DSDP. Initial Reports Deep Sea Drilling Project 42: 527-539.

14. CAMPBELL, P.G.C., STOKES, P.M. and GALLOWAY, J.N. (1983). Effects of atmospheric deposition on the geochemical cycling and biological availability of metals. Proc. Intern. Conf. Heavy Metals in the Environment, Heidelberg, pp. 760-763.

15. CHANEY, R.L., BROWN, J.C. and TIFFIN, L.O. (1972). Plant Physiol. 50: 208-221. Cit. (79).

16. CHAO, L.L. (1972). Selective dissolution of manganese oxides from soils and sediments with acidified hydroxylamine hydrochloride. Soil Sci. Soc. Amer. Proc. 36: 764-768.

17. CHESTER, R. and HUGHES, M.J. (1967). A chemical technique for the separation of ferromanganese minerals, carbonate minerals and adsorbed trace elements from pelagic sediments. Chem. Geol. 2: 249-262.

51. HAM, R.K., ANDERSON, M.A., STANFORTH, R. and STEGMANN, R. (1979). Background study on the development of a standard leaching test. EPA-600/2-79-109.

52. HANSEN, L.D. and FISHER, G.L. (1980). Elemental distribution in coal fly ash particles. Environ. Sci. Technol. 14: 1111-1117.

53. HARRISON, R.M., LAXEN, D.P.H. and WILSON, S.J. (1981). Chemical associations of lead, cadmium, copper, and zinc in street dusts and roadside soils. Environ. Sci. Technol. 15: 1378-1383.

54. HART, B.T. (1981). Trace metal complexing capacity of natural waters. A review. Environ. Technol. Letts. 2: 95-110.

55. HART, B.T. (1982). Uptake of trace metals by sediments and suspended particulates. A review. Hydrobiologia 91: 299-313.

56. HEATH, G.R. and DYMOND, J. (1977). Genesis and transformations of metalliferous sediments from the East Pacific Rise, Bauer Deep and Central Basin, Nasza Plate. Geol. Soc. Amer. Bull. 88: 723-733.

57. HESSLEIN, R. (1976). An in situ sampler for close interval pore water studies. Limnol. Oceanogr. 21: 912-914.

58. HIRST, D.M. and NICHOLLS, G.D. (1958). Techniques in sedimentary geochemistry. 1. Separation of the detrital and non-detrital fractions of limestone. J. Sediment. Petrol. 28: 461-468.

59. HOLMGREN, G.S. (1967). A rapid citrate-dithionite extractable iron procedure. Soil Sci. Soc. Amer. Proc. 31: 210-211.

60. HUNTER, K.A. (1980). Processes affecting particulate trace metals in the sea surface microlayer. Mar. Chem. 9: 49-70.

61. HUNTER, K.A. (1983). On the estuarine mixing of dissolved substances in relation to colloid stability and surface properties. Geochim. Cosmochim. Acta 47: 467-473.

62. International Atomic Energy Agency. (E.D. HESPE, 1971). Leach testing of immobilized radioactive waste solids. Atomic Energy Review 9: 195-207.

63. International Council for the Exploration of the Sea (ICES) (1983). Intercalibration study 1984.

64. JACKSON, M.L. (1958). Soil Chemical Analysis. Prentice-Hall, Englewood Cliffs, New Jersey, 498 p.

65. JENNE, E.A. (1977). Trace element sorption by sediments and soils - sites and processes. In: W.CHAPPELL and K.PETERSEN (eds.) Symposium on Molybdenum. Marcel Dekker, New York, pp. 425-553.

66. JENNE, E.A. and LUOMA, S.N. (1977). Forms of trace elements in soils, sediments, and associated waters: An overview of their determination and biological availability. In: R.WILDUNG and H.DRUCKER (eds.) Biological Implications of Metals in the Environment. Conf. 750929 NTIS Springfield VA, p. 110-143.

67. JENNE, E.A., KENNEDY, V.C., BURCHARD, J.M. and BALL, J.W. (1980). Sediment collection and processing for selective extraction and for total trace element analysis. In: R.A. BAKER (ed.) Sediments and Contaminants. Ann Arbor Sci. Publ., Vol. 2, pp. 169-191.

68. JONASSON, I.R. (1977). Geochemistry of sediment/water interactions of metals, including observations on availability. In: H. SHEAR and A.E.P. WATSON (eds.) The Fluvial Transport of Sediment-Associated Nutrients

and Contaminants. IJC/PLUARG, Windsor/Ont., pp. 255-271.

69. JONES, A.S.G. (1973). The concentration of copper, lead, zinc, and cadmium in shallow marine sediments, Cardigan Bay, Wales. Mar. Geol. 14: M1-M9.

70. KEENEY, D.R. and WALSH, L.M. (1977). Heavy metal availability in sewage sludge amended soils. Proc. Int. Conf. Heavy Metals in the Environment, Toronto 1975, II/1: 379-402.

71. KENNEDY, V.C., JONES, B.F. and ZELLWEGER, G.W. (1974). Filter pore-size effects on the analysis of Al, Fe, and Ti in water. Water Resources Res. 10: 785-790.

72. KENNEDY, V.C., JENNE, E.A. and BURCHARD, J.M. (1976). Backflushing filters for field processing of water samples prior to trace-element analysis. U.S. Geol. Survey, Open-File Rept. 76-126, pp. 1-12.

73. KENNEDY, V.C., ZELLWEGER, G.W. and AVANZINO, R.J. (1979). Variation in rain chemistry during storms at two sites in northern California. Water Resources Res. 15: 687-702.

74. KERSTEN, M. (1984). Unpublished data, Arbeitsbereich Umweltschutztechnik TU Hamburg-Harburg.

75. KEYSER, T.R., NATUSCH, D.F.S., EVANS, C.A.jr and LINTON, R.W. (1978). Characterizing the surface of environmental particles. Environ. Sci. Technol. 12: 768-773.

76. LAXEN, D.P.H. and SHOLKOVITZ, E.R. (1981). Adsorption (coprecipitation) of trace metals at natural concentrations on hydrous ferric oxide in lake water samples. Environ. Technol. Letts. 2: 561-569.

77. LEE, G.F. (1975). Role of hydrous metal oxides in the transport of heavy metals in the environment. In: P.A. KRENKEL (ed.) Heavy Metals in the Aquatic Environment. Pergamon, Oxford, pp. 137-147.

78. LERMAN, A. (1977). Migrational processes and chemical reactions in interstitial waters. In: E.D. GOLDBERG (ed.) The Sea. Wiley, New York, Vol. 6.

79 LINDSAY, W.L. (1974). Role of chelation in micronutrient availability. In: E.W. CARSON (ed.) The Plant Root and its Environment. Univ. Virginia Press, Charlotsville, pp. 507-524.

80. LINTON, R.W., NATUSCH, D.F.S., SOLOMON, R.L. and EVANS, C.A. jr. (1980) Physicochemical characterization of lead in urban dusts. A microanalytical approach to lead tracing. Environ. Sci. Technol. 14: 159-164.

81. LION, L.W., ALTMAN, R.S. and LECKIE, J.O. (1982). Trace-metal adsorption characteristics of estuarine particulate matter: Evaluation of contribution of Fe/Mn oxide and organic surface coatings. Environ. Sci. Technol. 16: 660-666.

82. LORING, D.H. and RANTALA, R.T. (1977). Geochemical analysis of marine sediments and suspended particulate matter. Fish. Mar. Serv. Tech. Rept. No. 700, 58 p.

83. LUOMA, S.M. and BRYAN, G.W. (1978). Factors controlling the availability of sediment-bound lead to the estuarine bivalve Scrobicularia plana. J. Mar. Biol. Assoc. U.K. 58, 793-802.

84. LUOMA, S.N. and BRYAN, G.W. (1979). Trace metal bioavailability:modelling chemical and biological interactions of sediment-bound zinc. In: E.A. JENNE (ed.) Chemical Modelling in Aqueous systems. Amer. Chem. Soc. Symp. Ser. 93: 577-609.

85. MALO, B.A. (1977). Partial extraction of metals from aquatic sediments. Environ. Sci. Technol. 11: 277-282.

86. MATTIGOD, S.V. and PAGE, A.L. (1983). Assessment of metal pollution in soils. In: I. THORNTON (ed.) Applied Environmental Geochemistry. Academic Press, London, pp. 355-394.

87. MAYER, L.M. (1976). Chemical water sampling in lakes and sediments with dialysis bags. Limnol. Oceanogr. 21: 909-911.

88. MEIGGS, T.O. (1980). The use of sediment analysis in forensic investigations and procedural requirements for such studies. In: R.A. BAKER (ed.) Contaminants and Sediments. Ann Arbor Sci. Publ., Vol 1:297-308.

89. MOORE, D.P. (2974). In: E.W. CARSON (ed.) The Plant Root and its Environment. Univ. Press of Virginia, Charlotsville, pp. 135-151. Cit.(79).

90. MOORE, R.M., BURTON, J.D., WILLIAMS, P.J. and YOUNG, M.L. (1979). The behaviour of dissolved organic material, iron and manganese in estuarine mixing. Geochim. Cosmochim. Acta 43: 919-926.

91. NATUSCH, D.F.S. and TAYLOR, D.R. (1980). Environmental effects of western coal combustion. Part IV - Chemical and physical characterization of coal fly ash. Final Report Contract No. R803950, U.S. EPA 600/3-80-093. 200 p.

92. NÜRNBERG, H.W. (1983). Investigations on heavy metal speciation in natural waters by voltammetric procedures. Fresenius Z. Anal. Chem. 316: 557-565.

93. PATCHINEELAM, S.R. (1975). Untersuchungen über die Hauptbindungsarten und die Mobilisierbarkeit von Schwermetallen in fluviatilen Sedimenten. Dissertation University Heidelberg, 136 p.

94. PFEIFFER, G., FÖRSTNER, U. and STOFFERS, P. (1982). Speciation of reducible metal compounds in pelagic sediments by chemical extraction. Senckenbergiana Maritima 14: 23-38.

95. PICKERING, W.F. (1981). Selective chemical extraction of soil components and bound metal species. CRC Critical Rev. Anal. Chem. Nov. 1981: 233-266.

96. PLANT, J.A. and RAISWELL, R. (1983). Principles of environmental geochemistry. In: I. THORNTON (ed.) Applied Environmental Geochemistry. Academic Press, London, pp. 1-39.

97. RAPIN, F. and FÖRSTNER, U. (1983). On the selectivity of various extractants used in sequential leaching techniques for particulate metal speciation. Proc. Int. Conf. Heavy Metals in the Environment, Heidelberg, pp. 1074-1077.

98. RENDELL, P.S., BATLEY, G.E. and CAMERON, A.J. (1980). Adsorption as a control of metal concentrations in sediment extracts. Environ. Sci. Technol. 14: 314-318.

99. REUTHER, R., WRIGHT, R.F. and FÖRSTNER, U. (1981). Distribution and chemical forms of heavy metals in sediment cores from two Norwegian lakes affected by acid precipitation. Proc. Int. Conf. Heavy Metals in the Environment, Amsterdam, pp. 318-321.

100. REUTHER, R., ANDERSSON, J.B. and FÖRSTNER, U. (1983). Effect of pH-changes on the distribution and chemical forms of heavy metals in sediment cores from Swedish lakes. Proc. Int. Conf. Heavy Metals in the Environment, Heidelberg, pp. 868-871.

101. ROSE, A.W., HAWKES, H.E. and WEBB, J.S. (1979). Geochemistry in Mineral Exploration. Academic Press, London, 657 p.

102. SAAR, R.A. and WEBER, J.H. (1982). Fulvic acid: Modifier of metal-ion chemistry. Environ. Sci. Technol. 16: 510A-517A.

103. SALOMONS, W. (1980). Adsorption processes and hydrodynamic conditions in estuaries. Environ. Technol. Letts. 1: 356-365.

104. SALOMONS, W. and FÖRSTNER, U. (1980). Trace metal analysis on polluted sediments. II. Evaluation of environmental impact. Environ. Technol. Letts. 1: 506-517.

105. SALOMONS, W. and FÖRSTNER, U. (1984). Metal in the Hydrocycle. Springer-Verlag, Berlin-Heidelberg-New York-Tokyo, 349 p.

106. SALOMONS, W. and MOOK, W.G. (1977). Trace metal concentrations in estuarine sediments: Mobilization, mixing or precipitation. Neth. J. Sea Res. 11: 199-209.

107. SALOMONS, W., VAN DRIEL, W., KERDIJK, H. and BOXMA, R. (1982). Help! Holland is plated by the Rhine (environmental problems associated with contaminated sediments). Effects of Waste Disposal on Groundwater. Proc. Exeter Symposium IHAS 139: 255-269.

108. SCHOER, J., NAGEL, U., EGGERSGLUESS, D. and FÖRSTNER, U. (1982). Metal contents in sediments from the Elbe, Weser and Ems Estuaries and from the German Bight (southeastern North Sea): Grain size effects. Mitt. Geol. Paläontol. Inst. Univ. Hamburg, SCOPE/UNEP Special Volume 52: 687-702.

109. SCHWERTMANN, U. (1964). Differenzierung der Eisenoxide des Bodens durch photochemische Extraktion mit saurer Ammoniumoxalat-Lösung. Z. Pflanzenernährung Düngem. Bodenk. 105: 194-202.

110. SHOLKOVITZ, E.R. (1978). The flocculation of dissolved Fe, Mn, Al, Cu, Ni, Co and Cd during estuarine mixing. Earth Planet. Sci. Lett. 41: 77-86.

111. SYMEONIDES, C. and McRAE, S.G. (1977). The assessment of plant-available cadmium in soils. J. Environ. Qual. 6: 120-123.

112. TADA, F. and SUZUKI, S. (1982). Adsorption and desorption of heavy metals in bottom mud of urban rivers. Water Res. 16: 1489-1494.

113. TESSIER, A., CAMPBELL, P.G.C. and BISSON, M. (1979). Sequential extraction procedure for the speciation of particulate trace metals. Anal. Chem. 51: 844-851.

114. THOMAS, R.L. (1972). The distribution of mercury in the sediment of Lake Ontario. Can. J. Earth Sci. 9: 636-651.

115. THOMAS, A.J. and MARTIN, J.M. (1982). Chemical composition of river suspended sediment: Yangtse, Mackenzie, Indus, Orinoco, Parana and French Rivers (Seine, Loire, Garonne, Dordogne, Rhône). Mitt. Geol. Paläontol. Inst. Univ. Hamburg, SCOPE/UNEP Special Volume 52: 555-564.

116. TIPPING, E. (1981). The adsorption of aquatic humic substances by iron hydroxides. Geochim. Cosmochim. Acta 45: 191-199.

117. TURNER, R.R., LOWRY, P., LEVIN, M., LINDBERG, S.E. and TAMURA, T. (1982). Leachability and Aqueous Speciation of Selected Trace Constituents of Coal Fly Ash. Final Report, Research Project 1061-1/EA-2588, Electric Power Research Institute, EPRI-EA-2588.

118. U.S. Environmental Protection Agency (1980). U.S. EPA Hazardous Waste and Consolidated Permit Regulations. Federal Register, Vol. 45, No. 98, Book 2.

119. U.S. Geological Survey (1972). Geological Survey Research. USGS Prof. Paper 800-A, p. 147. Cit. (67).

120. VOLKOV, I.I. and FOMINA, L.S. (1974). Influence of organic matter and processes of sulfide formation on distribution of some trace elements in deepwater sediments of Black Sea. In: E.T.DEGENS and R.A.ROSS (eds) The Black Sea - Geology, Chemistry and Biology. Amer. Assoc.Petrol. Geol. Memoir 20: 456-476.

121. WEBBER, M.D. and CORNEAU, D.G.M. (1977). Metal extractability from sludge-soil mixtures. Proc. Int. Symp. Heavy Metals in the Environment, Toronto, 1975, Vol. I: 205-225.

122. WILBER, W.G. and HUNTER, J.V. (1979). The impact of urbanization on distribution of heavy metals in bottom sediments of the Saddle River. Water Resources Bull. 15: 790-800.

123. ZIRINO, A. and HEALY, M.L. (1971). Voltammetric measurement of zinc in the Northeastern Tropical Pacific Ocean. Limnol. Oceanogr. 16: 773-778.

DISCUSSION

Chairman: R.D. DAVIS

Rapporteurs: G.C. PORTEOUS, T.M. ADAMS, J.R. SAUNDERS

The fixation of metals on sediments varied. Zn and Cd were less strongly fixed than Cu and Pb, possibly due to differences in hydrolysis constants. Leckie's laboratory (Stanford, USA) had done some work on this subject.

Mo and Se were more available under oxidising conditions in sediments of low organic matter content but in soils they tended to be more available in reducing conditions because of the presence of organic matter. The paper had been confined to sediment studies.

A good indicator of the stability of metal compounds in relation to pH value was obtained by determining the pH_{50} value at which 50 per cent of the metal was in solution and 50 per cent adsorbed.

In sequential extraction techniques the pH value of the extractant was more important than the solid:extractant ratio. Peaks in extractability of metals were dependent on both pH value and redox conditions.

Concentrated nitric acid was the preferred extractant for determining 'total' metal concentrations in sediments because it was associated with minimal problems of precipitation.

The concentration of metals in pore water was probably most closely related to their environmental activity but if the metals in pore water were complexed they would not necessarily be bioavailable. (Contributors to discussion : Sanders, Fleming, Kick, Webber, Sauerbeck, Tjell and Herms)

PRINCIPLES OF INVESTIGATIONS ON THE MOBILITY AND PLANT UPTAKE OF HEAVY METALS

L. KIEKENS and A. COTTENIE
Technical College for Chemistry, Textile and Agriculture, CTL, Gent
Laboratory Analytical and Agrochemistry, State University Gent Belgium

Summary

Different approaches for chemical soil analysis are discussed. In comparison with conventional extraction methods, the study of the mobility has the advantage that the nature of the element and the effect of soil properties are also reflected in the amounts of heavy metals solubilized at different pH values.
Very significant relationships have been observed between metal concentrations in ryegrass and the amounts of Zn, Cu, Cd and Ni mobilized at different pH values.
On the contrary, the correlations of plant concentrations and the exchangeable and complexed forms of metals as determined by respectively 1 N NH_4Ac pH = 7 and 0.2 N NaOH showed quite large variations, mainly in function of soil type and element species.
Determination of the mobility of heavy metals in soils may be considered as a "more universal" method for assessing the biological effects of soil enrichment with heavy metals.

1. INTRODUCTION

Chemical methods for the characterization of soil pollution are in many respects different from those used for the determination of the nutrient status of soils. Total analysis for example has little value when estimating the nutrient status of soils, since a more or less important fraction of the total element content may be present in a non-available form for plants. However, the total contents may be considered when a suspected soil enrichment has to be judged or when maximum acceptable values are fixed.

Numerous methods have been described for the determination of nutrient elements in soils. They generally consist in solubilizing fractions of elements which are considered as available, accessible, mobile or simply extractable.
Some extracting solutions are used for selective determination of one single element, while others aim at the simultaneous determination of a number of elements.
The choice of extracting solutions has often been based on empirism, rather than purely theoretical considerations, though some more theoretical aspects may have inspired their adoption. Their value is normally judged in function of more or less satisfying correlations between the extracted amounts and plant growth or nutrient uptake.
The analytical results are moreover influenced by the operating modalities such as soil/solution ratio, extraction time, soil type and chemical properties of the extracting solution.

As the biologically active fraction of heavy metals in soils mainly consists of their soluble, exchangeable and complexed forms, it is evident to base their extraction on the following phenomena : solubilization of components in the solid soil phase, ion exchange (desorption) and complexation. Thus an extracting solution offering most chances to provide the required correlation with plant uptake should meet the following criteria :

a) be sufficiently acid to solubilize a solid phase fraction which contributes to plant uptake,
b) contain a displacing ion in order to exchange a fraction of adsorbed ions,
c) have the property to extract soluble organomineral complexes.

Based upon these considerations and in view of planning international programmes, the FAO European Cooperative Network has proposed a reference method using 0.5 n NH_4Ac + 0.02 M EDTA pH 4.65 as extracting solution (6). However a single extraction only reflects the situation at a given moment and does not provide information about the effects to be expected over a longer lapse of time.

The extracted amounts represent the equilibrium concentration obtained by the contact of the solid soil phase with the extracting solution. Increasing the agressivity of the extracting solution also will increase the extracted amounts.

Viets (9) stated that soil constituents can stepwise be brought into solution, using extractants of increasing agressivity, starting with pure water. The final attack with strong acids is the last integration step to obtain the total amount. It is not known up to which pool plants are able to take up an element and the various plant species certainly act quite different with regard to the agressivity of their root system.

In recent years a number of sequential extraction procedures for heavy metals in soils, sludges and sediments have been proposed (1, 3, 4, 7, 8).

The main purpose of fractionation is to better understand the distribution of heavy metals over different forms. Various reagents have been proposed to extract fractions representing organic, soluble, exchangeable, adsorbed and precipitated forms.

The great diversity of extraction and fractionation procedures and consequently the difficulty of comparing results obtained by different methods have focussed the attention to develop "a more universal method" for assessing the availability of heavy metals in soils.

The actual concentration of heavy metals in the soil solution is governed by a number of competitive reactions such as precipitation and dissolution, complexation and decomplexation, adsorption and desorption. If one factor of such a dynamic system is changed, a shift of the global equilibrium will occur until a new equilibrium situation is reached.

Practically pH seems to be a very suitable parameter to induce equilibrium displacements in the soil-soil solution system, thus leading to a transfer of elements from the solid to the liquid soil phase or vice versa.

In fact, upon acidification of a soil-water suspension an increase of the concentration of heavy metals in the soil solution will occur, due to phenomena such as solubilization, decomposition of organomineral complexes and desorption.

For estimating the nutrient status of soils, a distinction has often been made between :

- nutrient intensity, which is the actual soluble amount of nutrients, and
- nutrient capacity, representing a fraction which may become plant available during crop growth.

This approach is related to the concept that biological activity and plant uptake of nutrients are mainly determined by their mobility in the growth medium.

Based on these considerations, a mobile fraction of an element may be defined as the sum of the actual amount present in the soil solution, corresponding with the above mentioned nutrient intensity, and an amount present in the solid phase which can be transferred to the liquid phase by changing the pH of the soil.
Thus the analytical determination of the mobility of heavy metals by stepwise acidification of a soil-water suspension will be more than a single extraction with any extracting agent.
The purpose of the present paper was to discuss some experimental results concerning the effect of the mobility of heavy metals in soils on their uptake by plants.

2. ANALYTICAL PROCEDURE (5)

For the experimental determination of the mobilization pattern, extraction at at least four pH steps is required.
A. Original pH of the soil-water suspension (pH soil)
B. pH 4.0
C. pH 2.0
D. pH 0.5
For each pH value a separate subsample is used.

A. Extraction at the original pH of the soil
Weigh 10 g of soil in a 150 ml beaker, add 50 ml H_2O, stir for 30 minutes. Leave overnight, stir again for 5 minutes and measure the pH (pH-soil). Centrifuge for 15 minutes at 3000 g and take an aliquot of the clear supernatant solution for analysis.

B. C and D (see above)
Weigh 10 g of soil in a 150 ml beaker, add 30 ml H_2O and adjust the pH of the suspension to the desired value by addition of controlled volumes (V ml) HNO_3 not exceeding 20 ml.
The HNO_3 concentration may vary from 0.1 to 4 N as required by the aimed pH decrease and the nature of the soil (buffering capacity).
For light and medium textured soils 0.1 to 2 N HNO_3 is normally sufficient. For heavy soils with a high buffering capacity or for soils containing carbonates, it is recommended to use 4 N HNO_3 for adjusting the pH to low values. If any doubt should exist about the origin and nature of the soil sample, it is desirable to determine previously by means of a titration curve, which HNO_3 concentration is to be used for adjusting the pH to low values.
The pH of the continuously stirred suspension is kept constant during 30 minutes. To do so an automatic titrator (pH stat apparatus) is most useful.
If not available, the pH can also be adjusted by adding HNO_3 from a burette and recording the pH by means of a pH meter. After equilibration, bring the final volume to 50 ml by adding (20 - V) ml of distilled water, V being the volume of HNO_3 added during equilibration. Centrifuge the soil suspension at 3000 g for 15 minutes and take an aliquot of the clear supernatant solution for analysis.

3. EXPERIMENTAL INFORMATION

3.1. Mobility of heavy metals in soils

Application of the proposed mobility test to a large number of polluted and non-polluted soils have led to the following observations :

- typical mobilization patterns are obtained for the different heavy metals. This is illustrated in fig. 1 showing the percentages of total Zn, Cu, Cd, Pb, Ni, Co and Cr mobilized in function of pH in a sludge amended soil sample (5). The mobilization curves show a similar pattern for the metal-pairs Zn-Cd, Pb-Cu and Co-Ni. Cadmium showed the highest mobility while the lowest mobility was observed for Cr.
  Cd and Zn tend to pass into solution at higher pH values (pH $\simeq$ 4) than is the case for the other metals. The different behaviour of the metals may be linked to their involvement in different mechanisms governing their chemical behaviour in the soil. Indeed, it is well known that Cu and Pb are mainly involved in reactions with organic soil constituents, while Zn and mainly Cd take part in adsorption-desorption reactions.
- Besides the nature of the element, the physico-chemical properties of the soil also affect the mobilization pattern. The influence of soil texture on the mobility of Zn in a light, medium and heavy textured soil containing respectively 78, 79 and 76 ppm total Zn, is shown in figure 2. Though the three soils contain almost the same amounts of total zinc, its mobilization in function of pH is quite different.
- The proposed mobility test allows to differentiate between soils with normal and increased levels of heavy metals.
  The results obtained in an interlaboratory comparison carried out by the members of WP2 of the CEC concerted action COST 68 ter in two soil samples indicated that the mobility of most heavy metals was greater in the soil sample containing the highest total contents. Furthermore this ringtest has also demonstrated that the mobility test could be used successfully by most of the participating laboratories (5).

Thus one may conclude that in comparison with convential extracting methods, the study of mobility has the advantage that the nature of the element and the effect of soil properties are reflected in the mobile fractions of heavy metals in soils.

3.2. Mobility and plant uptake

At the third International Symposium on Processing and Use of Sewage Sludge held in Brighton in 1983 results of the effect of the mobility of Zn, Cu, Cd and Ni in the soil on plant uptake were presented (2).
For this study equilibration and pot experiments were conducted, using two different soils, a sandy and a heavy clay soil, amended with four levels of Zn, Cu, Ni and Cd. Heavy metals were added in two forms, namely as sulphate salts and incorporated in sewage sludge. After an equilibration period of six months, mobility of heavy metals in all treated soils was determined. Then a pot experiment was conducted in the greenhouse with Italian ryegrass (Lolium perenne cv. S24) as test crop.
The amounts of heavy metals solubilized in function of pH showed a systematic difference between the salt and sludge addition form, and also between both soils. This is illustrated in fig. 3 and 4, showing the mobility of heavy metals in both soils enriched with respectively 205 ppm Zn, 55 ppm Cu, 10 ppm Ni and 5 ppm Cd as inorganic salts and incorporated in sludge.
It is clear that for Zn, Cd and Ni the sequence of decreasing mobility was the following : sandy soil, salt form > sandy soil, sludge form $\approx$ clay soil, salt form > clay soil, sludge. The mobility curves of Cu in both soils and for both forms showed very little differences.
These mobility patterns correspond very well with the uptake of heavy metals by ryegrass, as illustrated in figure 5, showing plant uptake of heavy metals for both soils and both forms relative to the uptake for the blanc treatment.

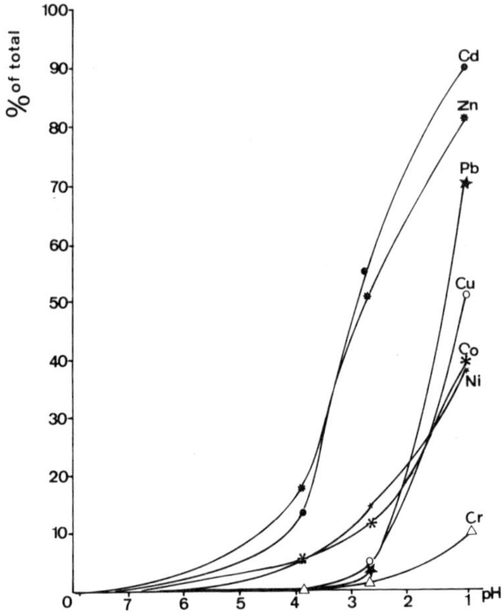

Fig. 1 - Percentage of total Zn, Cu, Cd, Ni, Co, Cr mobilized in function of pH in a sludge amended soil.

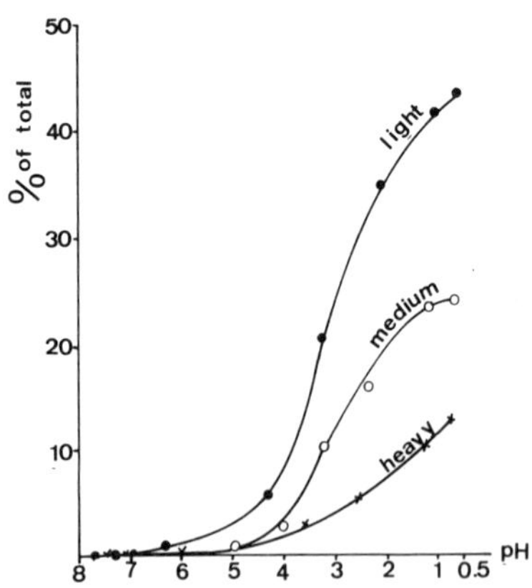

Fig. 2 - Percentage of total Zn solubilized in function of pH in 3 different textured soils

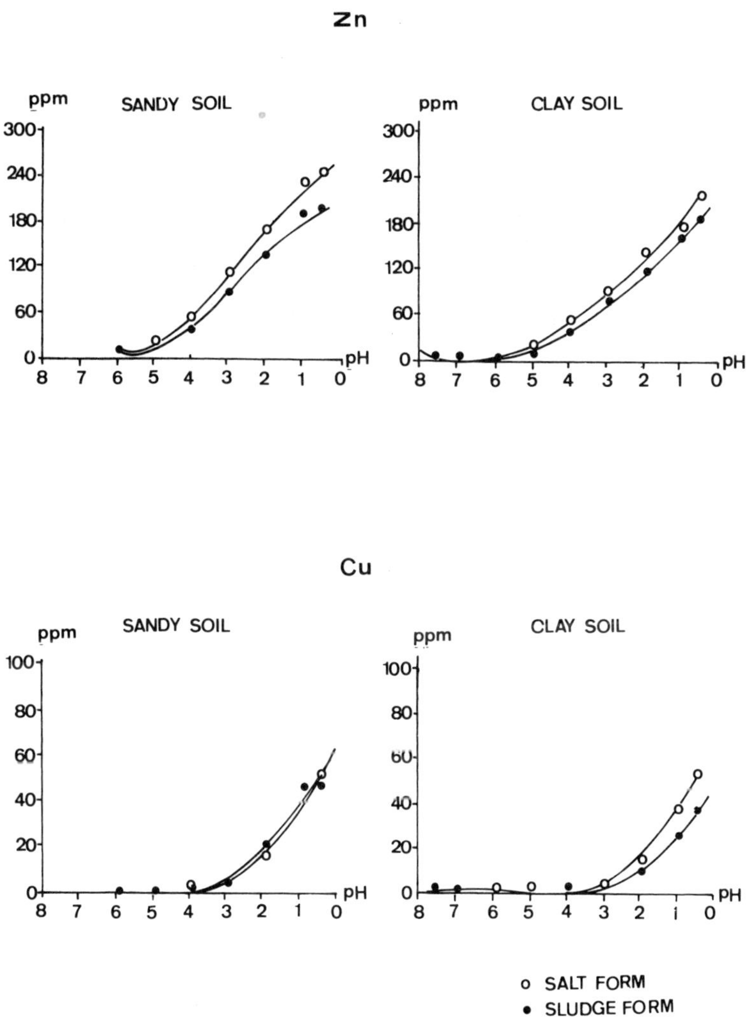

Fig. 3 - Quantities of Zn and Cu solubilized in function of pH for an addition level of 205 ppm Zn and 55 ppm Cu to the soil

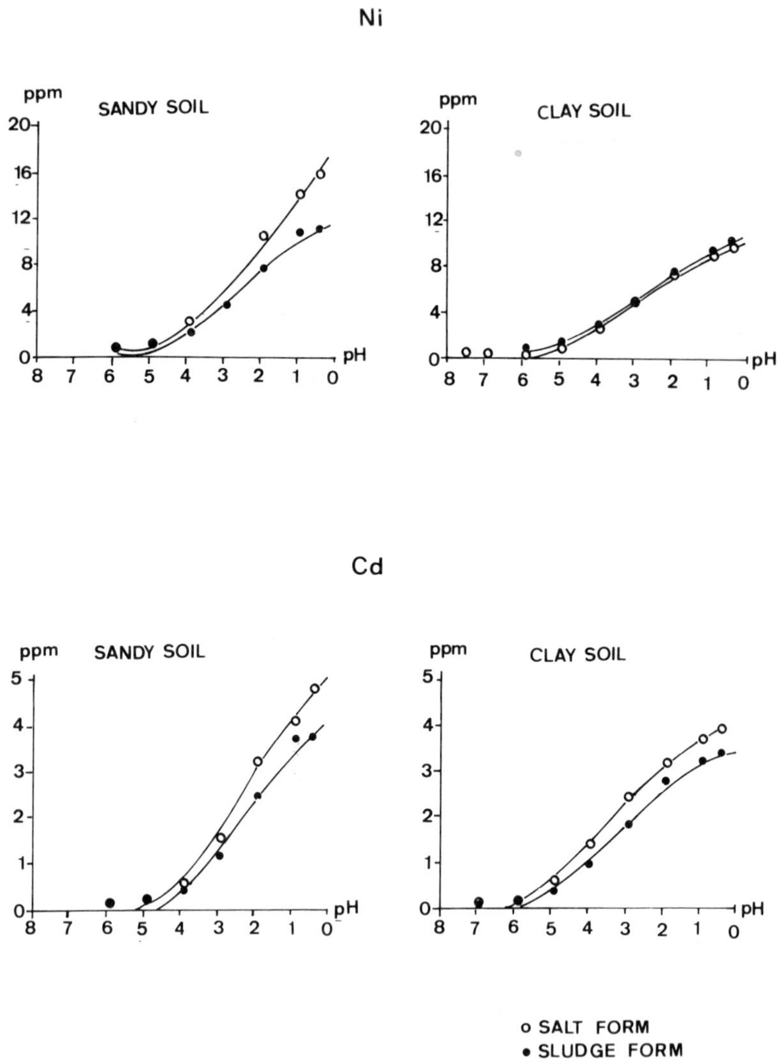

Fig. 4 - Quantities of Ni and Cd solubilized in function of pH for an addition level of 10 ppm Ni and 5 ppm Cd to the soil

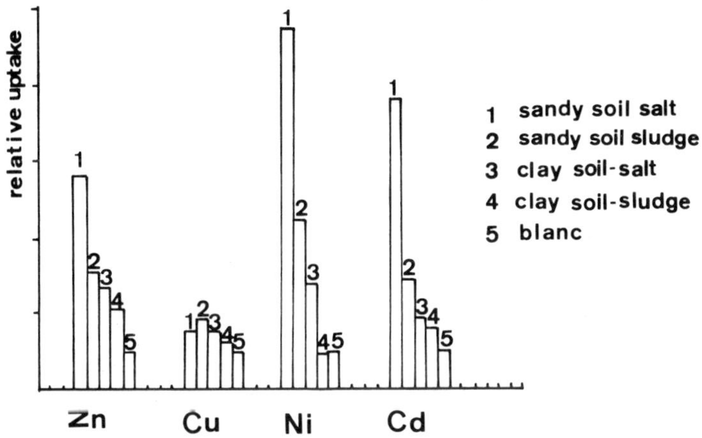

Fig. 5 - Relative metal uptake by ryegrass grown on a sandy soil and heavy clay soil amended with 205 ppm Zn, 55 ppm Cu, 10 ppm Ni and 5 ppm Cd in two forms (salts - sludge)

This figure clearly shows that metal uptake follows the same sequence as the mobility of heavy metals in both soils. Furthermore it seems that for the same soil and for the same metal form, plant uptake decreased in the following order Ni > Cd $\approx$ Zn > Cu.
Table I gives the correlation coefficients obtained from linear regression analysis across all treatments of plant concentrations of heavy metals with heavy metal content of the soil as determined by stepwise acidification and by other extractants.

Table I. Correlation coefficients obtained from linear regression analysis across all treatments of plant concentration of heavy metals with heavy metal content of the soil as determined by stepwise acidification of a soil-water suspension and with other extractants

Soil content	Plant concentration			
	Zn	Cu	Cd	Ni
1. Sandy soil				
pH 0.5	0.853**	0.814**	0.793**	0.759**
pH 1	0.837**	0.877***	0.795**	0.800**
pH 2	0.887***	0.900***	0.887***	0.862***
pH 3	0.907***	0.886***	0.943***	0.942***
pH soil	0.978***	0.639*	0.995***	0.976***
1 N NH_4Ac pH = 7	0.992***	0.780**	0.981***	0.990***
0.2 N NaOH	0.970***	0.972***	0.705*	0.697*
aqua regia	0.726*	0.863***	0.717*	0.692*
2. Heavy clay soil				
pH 0.5	0.979***	0.728*	0.900***	0.855***
pH 1	0.996***	0.754**	0.913***	0.907***
pH 2	0.995***	0.702*	0.912***	0.903***
pH 3	0.993***	0.666*	0.946***	0.897***
pH soil	0.553^{ns}	0.603*	0.706*	0.487^{ns}
1 N NH_4Ac pH = 7	0.759**	0.690*	0.958***	0.894***
0.2 N NaOH	0.944***	0.810**	0.706*	0.284^{ns}
aqua regia	0.974***	0.752**	0.902***	0.847***

*, **, *** indicate 0.05, 0.01, 0.001 levels of significance respectively
ns not significant

It seems that for both soils very significant correlations of plant concentrations with the amounts of heavy metals solubilized at different pH values were obtained. In most cases the correlation was improved by stepwise increasing the pH value from 0.5 to 3. At the actual soil pH the calculated correlation coefficients were distinctly different for both soils. This is also the case for 1 N NH_4Ac pH = 7 and 0.2 N NaOH, solubilizing fractions which are representative for respectively the easily exchangeable and complexed forms of heavy metals. The correlations of plant concentrations with the amounts extracted by both solutions show a quite large variation, mainly depending on soil type and the nature of the element.

The correlation coefficients for aqua regia extractable contents and the amounts solubilized at pH = 0.5 were very similar for both soils and for all metals studied.
Summarizing one may conclude that the proposed mobility test for heavy metals in soils may be considered as a "more universal" method for assessing the biological effect of heavy metals in soils. Stepwise acidification of a soil-water suspension releases amounts of heavy metals which are indicative for as well the actual available quantities as those which may become available in the course of time.

REFERENCES

1. BADRI, M.A. and ASTON, S.R. (1981). Fractionation of heavy metals in estuarine sediments. Proc. Intern. Conf. Heavy Metals in the Environment, Amsterdam, pp 705-708
2. COTTENIE, A., KIEKENS, L. and VAN LANDSCHOOT, G. (1983). Problems on the mobility and predictability of heavy metal uptake by plants. Proc. 3rd Int. Symposium Processing and Use of Sewage Sludge. Brighton
3. FORSTNER, U., CALMANO, W., CONRADT, K., JAKSCH, H., SCHIMKUS, C. and SCHOER, J. (1981). Chemical speciation of heavy metals in solid waste materials by sequential extraction. Proc. Intern. Conf. Heavy metals in the Environment. Amsterdam, pp 698-704
4. KIEKENS, L. and COTTENIE, A. (1983). Characterization of chemical and biological activity of heavy metals in the soil. Proc. Int. Conf. Heavy Metals in the Environment. Heidelberg, pp 657-661
5. KIEKENS, L. and COTTENIE, A. (1983). Report on the results of the interlaboratory comparison : Determination of the mobility of heavy metals in soils. Proc. 3rd Int. Symposium Processing and Use of Sewage Sludge, Brighton
6. LAKANEN, E. and ERVIO, R. (1971). A comparison of eight extractants for the determination of plant available micronutrients in soils. Acta Agr. Fenn., 123, 223-232
7. McLAREN, R.G. and CRAWFORD, D.V. (1973). Studies on soil copper. I. The fractionation of copper in soils. J. Soil Sci. 24, 172-181
8. TESSIER, A., CAMPBELL, P.G.C. and BISSON, M. (1979). Sequential extraction procedure for the speciation of particulate trace metals. Anal. Chem. 51, 844-851
9. VIETS, F. (1962). Chemistry and availability of micronutrients in soils. Agric. and Food Chemistry, 10, 174.

REASONS TO USE NEUTRAL SALT SOLUTIONS TO ASSESS THE METAL IMPACT ON PLANT AND SOILS

H. Häni and S. Gupta
Swiss Federal Research Station for Agricultural Chemistry and Hygiene of Environment

Summary

For polluted soils it is evident to use neutral salt solutions in order to determine metal fractions which are of some meaning for biological effects. The reasons are as follows:

- Almost uniform soil contents (soluble in 0.1 M $NaNO_3$) are found independently of soil properties to give defined biological effects. These values for zinc, cadmium and copper are 1, 0.06 and 1.5 ppm, respectively.
- The amounts extracted from soils by neutral salt solutions correspond fairly well to the amounts taken up by plants.
- At soil pH values lower than 6, cadmium and zinc are predominantly bound unspecifically. Extractants with pH values lower than 6 are, therefore, of little use to assess the bioavailability of metals in soil.

1. INTRODUCTION

The total content of heavy metals in soil is often used to express the degree of its contamination. However, this total amount is of little value to judge biological effects (e.g. on plants or microorganisms). It is the purpose of the present paper to show why neutral salt solutions are much better suited to assess the bioavailability of metals in polluted soils.

2. RESULTS AND DISCUSSION

In several experiments it was found that neutral salt solutions have important advantages compared with often used stronger extractants. In the following these advantages are discussed in detail.

2.1 Soil properties

In a multiple regression analysis with 36 typical Swiss soils it could be shown that the cation exchange capacity (CEC) is significantly correlated with the content of humus and clay. The iron- and aluminium-oxides seem to be much less important in influencing the cation exchange capacity:

$$CEC = 4.67 + 0.43(\text{clay}) + 2.2\ (\text{humus}) - 0.011(\text{Fe-oxide}) + 0.033(\text{Al-oxide})$$

$n = 36$

$r^2 = 0.83$

Furthermore, the metal content solubilized by a neutral salt solution is correlated with pH, CEC and total metal content of soil (1).

According to these correlations the soluble metal content is mainly governed by the pH value and the CEC of soil, whereby the pH is by far the most important factor. It is, therefore, understandable that hardly no distinction between soils of different metal binding capacity is possible when solvents which change the soil pH during the extraction procedure are chosen.

These purely soil chemical considerations are largely confirmed by experiments with plants and microorganisms. As can be seen from table 1, defined plant contents of the metals zinc, cadmium and copper are related to almost uniform $NaNO_3$-soluble soil contents, whereas the metal contents extracted by NH_4OAc (pH 4.8) and complexing agents differ widely from soil to soil. The results obtained with these solvents do not give more information than the total content, which means that each soil, depending on its properties, has its own value which leads to a defined plant content. - The results of the standardized cadmium pot trial are given as a further example for the good relation between $NaNO_3$-soluble soil cadmium and plant content (s. fig. 1).

From table 1 it is further evident that DTPA as a complexing agent of pH 7.3 dissolves the same amount of zinc as NH_4OAc of pH 4.8. In the case of copper, DTPA dissolves even more than NH_4OAc, which could be explained by the fact that copper ist preferentially bound to organic matter.

Table 1 Figures from glasshouse experiments: Ranges of heavy metal contents (total and soluble) which lead to phyto- (zinc and copper) or zootoxic concentrations in a given test plant in different soils (3). Zn, Cu: red clover (Trifolium pratense L.); Cd: rye grass (Lolium multiflorum Lam.).

Element	Total soil content (ppm)	Soluble soil content (ppm)				Plant content (ppm)
		NH_4OAc pH 4.8	NH_4OAc + EDTA pH 4.65	DTPA pH 7.3	$NaNO_3$ 0.1 M	
Zn	70-218	7-30	-	7-30	0.8-1.3	200
Cd	0.8-4	-	0.3-1.2	-	0.06	1
Cu	83-127	3.5-12	-	6.6-32	1-2	20

What has been observed for plants is also true for soil microorganisms. As an example, the effect of cadmium on the carbon mineralization is shown in table 2. The addition of sewage sludge to soil A drastically reduces the $NaNO_3$-soluble content of cadmium, which is reflected in a smaller reduction of the carbon mineralization.

Table 2 Figures from a glasshouse experiment:
Relation between cadmium in the soil solution (0.1 M $NaNO_3$) and the carbon mineralization (in %) (2).

Dose of Cd (ppm)	Dose of sewage sludge (g dry matter/ kg soil)	Soil A pH	Soil B pH	Cd in 0.1 M $NaNO_3$ (ppm)	% C mineralization
16	0	5.7		0.610	59
16	18	-		0.050	84
16	18		7.0	0.005	100

2.2 Comparison of metals extracted from soils with plant uptake

It is often observed that complexing agents and acid solvents extract more than 70 % of the metal amount which has totally been added to the soil. This may be another argument against the use of strong extractants because the cadmium figures in table 3 demonstrate that the amount taken up by maize is only about 5 per cent of the amount extracted.

Table 3 Figures from a glasshouse experiment:
Comparison of the cadmium uptake by maize with the amount extracted from soil with different solvents (2).

Cd addition (mg/pot)	Cd uptake (% of totally added)	Soluble Cd (% of totally added) $NaNO_3$	NH_4OAc	NH_4OAc + EDTA
23.3	4.0	8	69	75
47.0	3.0	8	71	78
95.0	3.0	9	71	80

2.3 pH dependency of the metal solubility in soils

The effect of the pH on the solubility of zinc and cadmium was studied in soils contaminated by the immissions of a smelter. After the addition of nitric acid and an equilibration time of 3 hours metal concentrations and pH were measured in the filtrate. The original pH of the soils (8.4-7.3) was lowered to 5 by this procedure. The cation exchange capacity of the soils differed from 10-30 meq/100 g.

The straight line shown for zinc in fig. 2 (c) has a slope of - 2.0, which means that one zinc ion is released by the addition of two protons. Furthermore, the solubility of zinc is increased by a factor of 100 when the pH is lowered by 1. The shift of this line to higher concentrations compared with the line b (Lindsay) may be due to the high total content of zinc in the investigated soils.

The results for cadmium are more difficult to interpret (s.fig.3). In no case a slope of - 2.0 is obtained. The fact that line b (soil contaminated by sewage sludge) is almost as flat as line a (uncontaminated soils) may be explained by a much greater influence of organic matter in binding the cadmium, thus it is rapidly trapped by the organic matter and no longer accessible to protons.

Coming back to the uniform soil contents ($NaNO_3$ values) mentioned under 2.1 it is interesting to note that these values are reached for both elements cadmium (s. table 4) and zinc (s. fig. 2) at pH of 6. The decrease of pH to 6 (due to fertilization, leaching of lime and nutrients, acid rain, etc.) in polluted soils is, therefore, undesired as the bioavailability of zinc and cadmium may rise to toxic levels. According to Scheffer / Schachtschabel (1982) the greatest part of cadmium and zinc is unspecifically bound at pH values lower than 6. In this form cadmium and zinc are easily exchangeable against alkaline earth ions. This is at the same time the third evidence why extractants with pH values lower than 6 are of little use to assess the bioavailability of metals in soil.

Table 4 Soluble contents of cadmium in polluted soils at pH around 6

Soil pH	Soil CEC	Soluble Cd (ppm)	pH
7.8	31.5	0.06	5.7
7.4	8.5	0.07	5.8
7.7	31.5	0.09	5.9
8.0	25.5	0.08	6.0
7.5	28.5	0.04	5.8
7.4	21.0	0.06	5.9
8.4	13.0	0.1	6.0

Due to its lower solubility, critical copper contents are not to be expected at soil pH values higher than 5 (s. fig. 2).

REFERENCES

1. GUPTA, S., HAENI, H. and SCHINDLER, P. (1980). Mobilisierung und Immobilisierung von Metallen durch die organische Bodensubstanz. Interner Forschungsbericht FAC, Liebefeld, 48 p.

2. GUPTA, S., HAENI, H. and SIEGENTHALER, A. (1983). Bedeutung von Gesamt- und biorelevanten Schwermetallgehalten zur Festlegung von Richtwerten in Böden. Schweiz. Landw. Fo. 22, 177-186

3. HAENI, H. and GUPTA, S. (1982). Total and biorelevant heavy metal contents and their usefulness in establishing limiting values in soils, Comm. of the European Communities, Proc. of a Workshop held at Stevenage, May 25-26, D. Reidel Publishing Company, 121-129.

4. HAENI, H. and GUPTA, S. (1983). Second report on the standardized Cd-pot experiment, EEC Concerted Action (COST 68), WP5 (83)2, 1-20

5. LINDSAY, W.L. (1981). In: Chemistry in the Soil Environment, ASA Special Publication Number 40, American Society of Agronomy and Soil Science Society of America, Madison, Wisconsin.

6. SOON, Y.K. (1981). Solubility and sorption of cadmium in soils amended with sewage sludge. Journal of Soil Science, 32, 85-95.

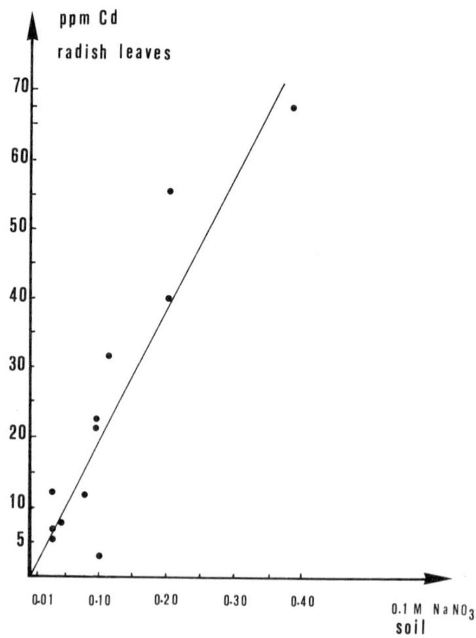

Fig. 1 - Standardized Cd-Pot Experiment 80/81 (4), Plant content of Cd in relation to 0.1 M $NaNO_3$-soluble soil Cd (Data from 12 different soils)

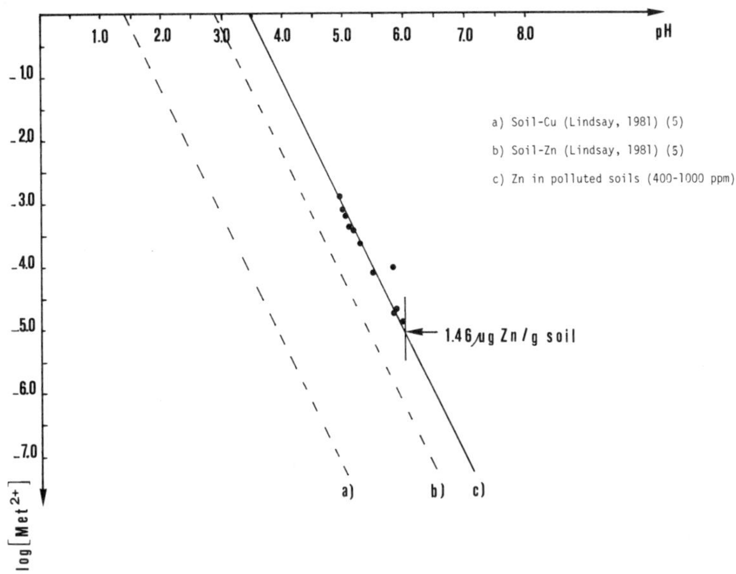

Fig. 2 - Solubility of zinc and copper in soils as a function of pH (slope s = - 2.0)

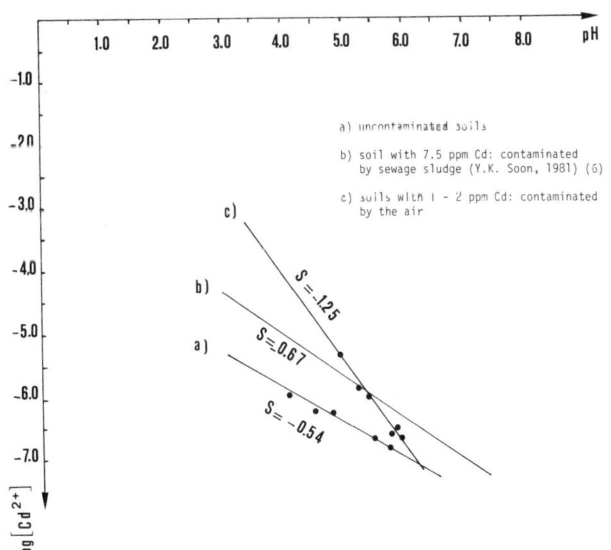

Fig. 3 - Solubility of cadmium in soils as a function of pH

DISCUSSION

Chairman: R.D. DAVIS

Rapporteurs: G.D. PORTEOUS, T.M. ADAMS, J.R. SAUNDERS

(Papers from L. KIEKENS and A. COTTENIE, and H. HÄNI and S. GUPTA)

In evaluating the efficacy with which chemical extractants measured bioavailability it was accepted that it was necessary to standardise on a single indicator plant (often ryegrass *Lolium* sp.) because plant concentrations of metals varied amongst different species. Spinach was a good indicator plant if high concentrations of metals (especially Cd) in plant tissue were wanted. But in the evaluation of extractants, effects of plant type had to be isolated to one side.

Extraction with sodium nitrate appeared to be very promising as a chemical method for measuring the bioavailability of cadmium in soil and the preparation of results of further evaluation tests on a wide range of soils was in progress. Results so far suggested that like the plant, sodium nitrate extraction was sensitive to soil conditions known to affect bioavailability such as cation exchange capacity and pH value. The progressive acidification technique was more drastic and likely to be less sensitive to subtle differences in soil conditions which were known to affect bioavailability profoundly. Further work was planned to compare the mobilisation profiles of the progressive acidification method with bioavailability to plants. Preliminary indications were that the concentration of metal extracted at pH value 3.0 correlated reasonably well with plant uptake. It was possible that the progressive acidification method was best regarded as a research tool for potentially identifying forms of metals in soils and how stable these were likely to be in the long-term following the effects of acid rain for instance. The progressive acidification procedure was considered to be laborious at least for routine purposes. On the other hand it had been used to evaluate several metals whilst work with sodium nitrate had focussed on Cd although the method showed promise for Cu and Zn. Like plants, sodium nitrate extracted only a small proportion of the total metal but this could pose analytical problems. For instance, the reliable detection limit for determining Cd in sodium nitrate extracts was about 0.06 mg/l which corresponded to about 2-3 ppm total Cd in soil. Therefore it would be difficult to use sodium nitrate to evaluate the bioavailability of Cd in uncontaminated soils where the total Cd content might be <1 ppm. The use of solvent extraction might be a way round this problem.

Before these methods could be recommended for regulatory purposes it would be necessary to evaluate analytical and sampling errors and to extend validation work over a wide range of soils contaminated by metals originating from various sources in addition to sewage sludge. Improved results might be obtained by grouping soils according for instance to their pH value or texture.

(Contributors to discussion : Fleming, Forstner, Hardwick, Herms, Davis, Kick, Muntau, Sanders, Sauerbeck, Tjell, Webber and Worthington)

EVALUATION OF CHEMICAL METHODS FOR ASSESSING THE Cd AND Zn AVAILABILITY FROM DIFFERENT SOILS AND SOURCES

D. R. SAUERBECK and P. STYPEREK*)
Institute of Plant Nutrition and Soil Science,
Federal Research Centre of Agriculture Braunschweig-Völkenrode (FAL),
Federal Republic of Germany

Summary

The uptake by plants has been compared from different soils and sources under both pot and field conditions with total and soluble contents of Cd and Zn in soils. Extractants with strong solubilization capacities and total contents did not, or not satisfactorily, indicate the experimentally found differences in availability. Neutral salt solutions like 0.1 M $NaNO_3$ and 0.05 or 0.1 M $CaCl_2$, on the other hand, both reflected the actual availabilities reasonably well. The extraction with $NaNO_3$, however, dissolved very little Cd from neutral soils, so that contamination degrees below 2 - 4 ppm total Cd were hardly detectable.
In contrast to this, the extraction with $CaCl_2$ turned out to be a successful tool, which did not only reflect the soil type dependent differences in availability but even the modifying influence of the particular source and the soil pH. A remarkably close correlation between plant uptake and $CaCl_2$ extractability has been ascertained at least for such plants and plant parts which absorbed significant trace element quantities. Hence, $CaCl_2$ seems a promising extractant in order to predict the actual Cd and Zn availability from different soils.

1. INTRODUCTION

It is well known that the transfer of heavy metals from soils to plants depends on the metal and plant species concerned and on the particular properties of the soil on which these plants grow. Of the trace metals relevant from the health and environmental point of view, Cd and Zn are the most mobile ones in regular soils. Moreover, they are frequently present in town waste and sewage sludge, and for this reason their uptake by crop plants from native and waste-treated soils is of considerable concern (1).

In order to control the sludge and town waste utilization in agriculture, several governments in Europe have already fixed maximum permissible heavy metal contents in sludges and soils, so as to maintain the environmental risk within acceptable limits (2, 3). These guidelines, however, do not take into consideration the paramount influence of individual soil properties and of the chemical form in which the metals occur. Hence, such administrative regulations based on total contents are frequently criticized, because on some soils they are far too strict, whereas on others the same limits can be far too high (4, 5, 6). This calls for additional criteria which enable the individual situations to be assessed, and for this purpose the potential of chemical soil test procedures needs to be evaluated.

*) under the cooperation of H. Kuntze, H. Kick, C. Förster, and C. Busch, Bonn and Bremen, Federal Republic of Germany

2. MATERIALS AND METHODS

Facing this situation, working party 5 of the EEC Coordinated Action COST 68 decided in 1980 to start a joint standardized pot experiment in order to study the effects of cadmium in soils with differing metal loading capacities. The general approaches and main objectives are summarized in Table I, of which the correlation between chemical extractability and plant availability is the main subject of this presentation. Some preliminary evaluations have already been presented by the Swiss coordinator of this joint experiment (4, 7), but did not yet consider the many additional facets which were investigated by the three participating institutions in Germany.

Table I: Standardized pot trial to study Cd-effects on different plants grown on soils with differing metal loading capacities

Main objectives:	A. Critical levels for different soils B. Toxicity levels for different plants C. Correlation between chemical extractability and plant availability of Cd in soils
Cadmium treatment:	0 - 1 - 2 - 4 - 8 (16) ppm Cd in soil
Kind of treatment:	Liquid sewage sludge (14 % d. m.) enriched with $CdSO_4$ after 14 d anaerobic incubation at 35 °C, 100 g d. m./pot
German participants:	Braunschweig (3 soils) Bonn (2 soils) Bremen (3 soils)

As Table II shows, these German pot experiments did not only use an artificially enriched sewage sludge but also soluble $CdSO_4$, and included a whole series of additional variations in order to study their influence on the Cd availability. Out of these many treatments only the liming and acidification effects will be dealt with, and to some extent also the deviating results obtained if an old highly contaminated sludge was used.

Table II: Cadmium sources and experimental treatments

Greenhouse experiments

Braunschweig	Bonn	Bremen
Cd-enriched sludge	Cd-enriched sludge	Cd-enriched sludge
pH-variation	pH-variation	
- liming	- liming	- liming
- acidification		
$MnSO_4$ addition	N-variation	moisture variation
org.matter addition	K-variation	Fe-variation
Mineral $CdSO_4$	Mineral $CdSO_4$	
Cd + Zn contam. dredged sludge		

The chemical characteristics of the 8 soils used for these pot experiments are shown in Table III. As can be seen from these data a reasonably wide spectrum of different soil properties was covered. The experimental crops of 1981 (spinach, bush beans and radish) corresponded to what was agreed upon internationally in working party 5. They were followed in 1982 by some other vegetables (lettuce, carrots and corn salad) whereas in 1983 the initial crop series was repeated in order to see whether the cadmium availability dropped as a function of time. Ryegrass was grown throughout the whole period (6 cuts per year), and a few more plants were tried (but not mentioned here) in some special cases.

Table III: Analytical characterization of soils

Soil type		%clay	%silt	pH	$\%CaCO_3$	% C	C/N	CEC	Cd_t
Phaeozem	Braunschweig	21	74	7.1	1.1	1.7	11	22.2	0.5
Luvisol	Braunschweig	15	82	7.3	1.3	1.2	10	16.4	0.5
Cambisol	Braunschweig	8	57	5.6	-	0.8	10	9.3	0.5
Luvisol	Bonn	16	77	5.5	-	1.0	11	11.2	0.5
Cambisol	Bonn	23	60	6.3	-	2.3	13	16.9	0.8
Areno-Histosol	Bremen	1	19	5.0	-	3.8	35	10.8	0.5
Gley-Podzol	Bremen	6	22	5.8	-	3.8	15	41.1	0.4
Podzol	Bremen	0	12	5.3	-	2.6	29	11.4	0.4

Regarding to the chemical extraction of soil-Cd it was decided for the EEC coordinated pot experiments to compare the progressive acidification method of Cottenie et al. (8, 9), the NH_4Ac-EDTA mixture of Lakanen and Ervioe (10, 11, 12), and the extraction of exchangeable Cd with $NaNO_3$ as suggested by Gupta et al. (13). After some promising tests done at Braunschweig, the German participants also included $CaCl_2$ (14, 15) as an additional extractant. The concentrations and soil/solution-ratios used are represented in Table IV.

Table IV: Extraction methods compared

- progressive acidification acc. to Cottenie (1 : 5)
- 0.025 M NH_4Ac + 0.02 M EDTA (1 : 5)
- 0.10 M $NaNO_3$ (1 : 2.5)
- 0.05 M $CaCl_2$ (1 : 2.5)
- 0.10 M $CaCl_2$ (1 : 2.5)

3. RESULTS AND DISCUSSION

Similar to the findings of other investigators (16, 17, 18, 19), growth and dry matter yields of the different crops as a function of soil Cd contents depended strongly on soil properties and on the plant species used. Critical soil cadmium concentrations varied from less than 2 up to more than 4 ppm Cd for light acid sands and for loam soils respectively, but ryegrass did even stand more than 8 ppm Cd without growth depressions. The cadmium from the artificially enriched sludge and from $CdSO_4$ turned out to be much more detrimental than the same concentrations derived from a several years old dried sludge material (6, 20).

Figure I is a typical example of the species-dependent plant Cd contents, confirming that leafy vegetables are strong accumulators for heavy metals, whereas grasses and legumes can discriminate against Cd especially on its way to their fruits and grains (16, 21, 22, 23, 24, 25, 26).

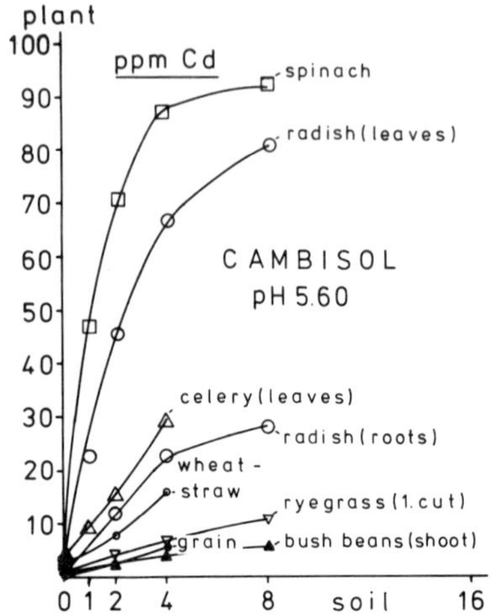

Figure I: Influence of increasing Cd-levels as enriched sewage sludge in soil on the Cd-content of different crop plants

Figure II shows the characteristic soil differences in Cd availability, indicating that the Cd contents of plants from acid soils can be several times higher than those from neutral or calcareous ones (5, 25, 27, 28, 29). This emphasizes the need for developing chemical methods to predict the actual Cd availability from different soils, which would be even more interesting if the behaviour of another metal of environmental concern like Zn could be reflected at the same time.

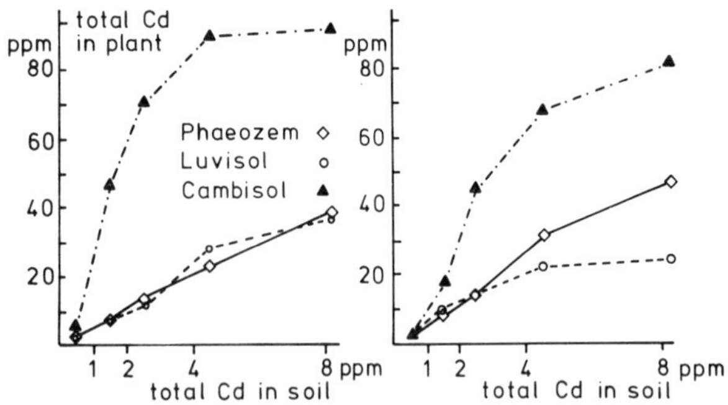

Figure II: Cd-content of spinach (left) and radish (right) versus total Cd-content in 3 different soils

Taking the results from Braunschweig as an example, Figure III indicates that the sludge-treated phaeozem (1) and the luvisol (2 a), although quite different in their pedological properties (Table III), did not differ very much in their amounts of Cd mobilized if both soils were adjusted to certain pH values by means of HNO_3 according to the step-wise acidification method of Cottenie et al. (8) for one hour. This most similar extraction behaviour contradicts with the actual differences in the plant availability from these two soils as tabulated for the Cd contents of radish plant leaves. Only the acid cambisol (3 a) exhibited a significantly higher Cd solubility which corresponded with the actual plant Cd contents.

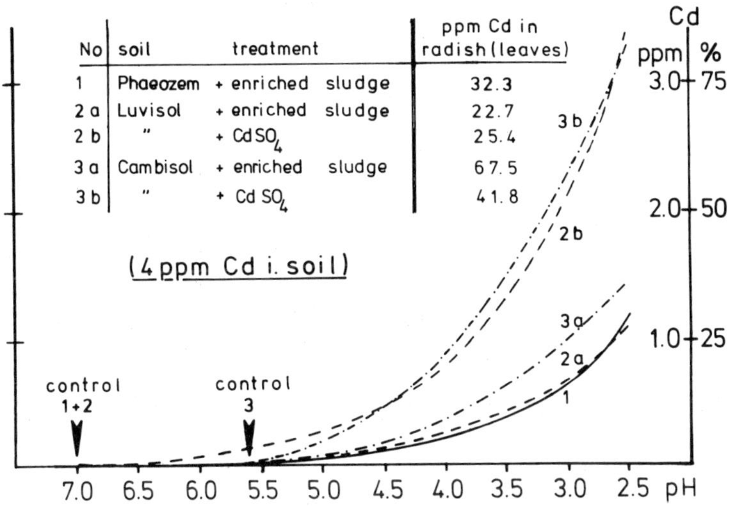

Figure III: pH-dependent dissolution of Cd (acc. to Cottenie, 1982) and its availability to radish plants from 2 sources on 3 different soils

No correspondence at all was, however, obtained if the soils 2 b and 3 b were treated with $CdSO_4$ instead of Cd enriched sludge. The Cd fraction dissolved from the $CdSO_4$ by the Cottenie method was in both soils considerably higher than that from the sludge-borne Cd, although the plant uptake did not differ very much or was even less from the $CdSO_4$ in the cambisol than that from the enriched sludge.

Even more disappointing were the results shown in Figure IV for the comparison of 2 soils receiving equal amounts of Cd from either artificially enriched or from old contaminated sludge. The Cd dissolution curves from both sources suggest an even higher availability from the dredged sludge material, but the plant Cd contents indicate exactly the opposite. Somewhat surprising is also the fact that liming which is known to reduce the plant uptake of Cd did not decrease but rather enhance its solubility in the soil suspensions adjusted to defined pH values. Our conclusion from these experimental findings is that the Cottenie method of progressive acidification is useful for principal investigations only (30, 31), but does not sufficiently characterize the actual plant availability of the Cd, not to speak of its considerable time and labour requirements which prevent its widespread application anyway.

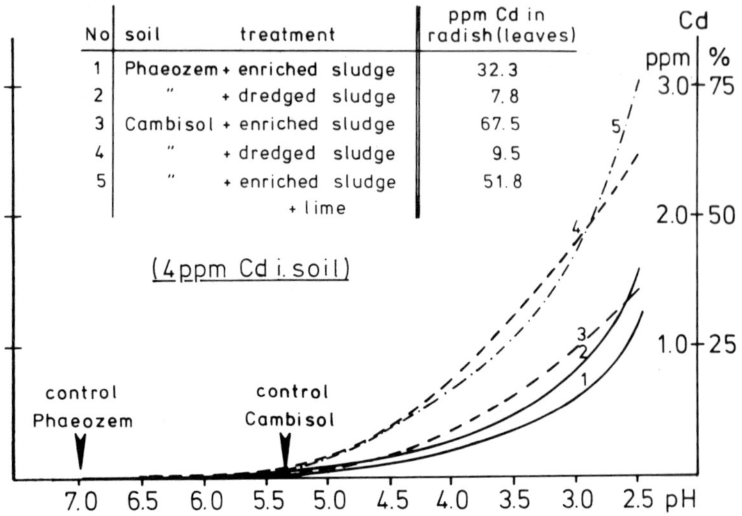

Figure IV: pH-dependent dissolution of Cd (acc. to Cottenie, 1982) and its availability to radish plants from 2 sludges on 2 different soils

As fas as the other internationally suggested extractant NH_4Ac-EDTA (10) is concerned, Figure V shows that irrespective of the soil properties

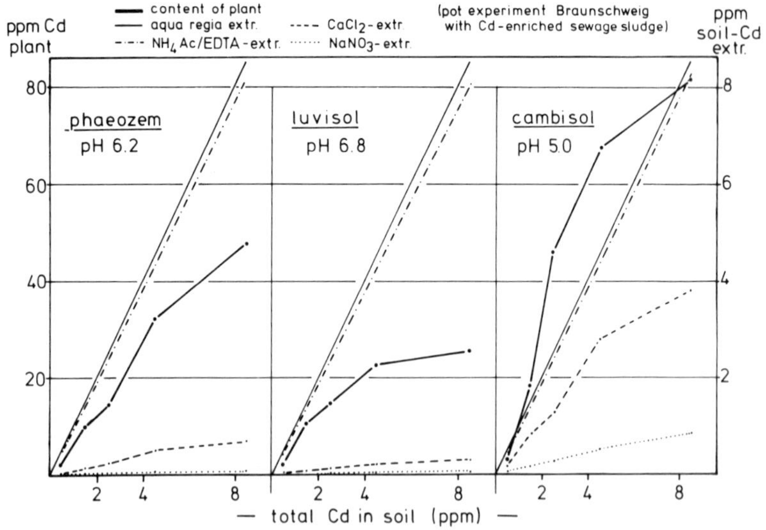

Figure V: Cadmium contents of radish leaves on 3 different soils, and amount of soil-Cd dissolved by various extractants

this mixture dissolved almost all the Cd present, and was only slightly less effective than the aqua regia extraction. As a consequence of this strong dissolution capacity almost no difference was found between soils, although the plants on these soils differed greatly in their Cd uptake. This result was fairly common in all cases tested, and therefore this particular method cannot be considered suitable for assessing the Cd status of German soils.

This may be quite different for other heavy metals like Cr and Pb, but as far as the pH-dependent behaviour and availability of Cd is concerned, the Swiss colleagues Häni and Gupta (4, 7, 32, 33) were certainly right when they suggested neutral salt solutions instead. Just like the Cd uptake by plants, the amounts which were dissolved by $NaNO_3$ differ greatly from soil to soil (see Figure V), and these differences in plant uptake and chemical extractability are obviously correlated. However, the amounts dissolved by $NaNO_3$ from neutral soils are often very small, so that even with graphite furnace techniques the soluble fractions at less than 2 ppm total Cd in most soils cannot be ascertained. Only on sandy or acid soils the $NaNO_3$ extractant dissolved enough Cd even at low contamination rates, but on other soils extremely sensitive determination methods would be required.

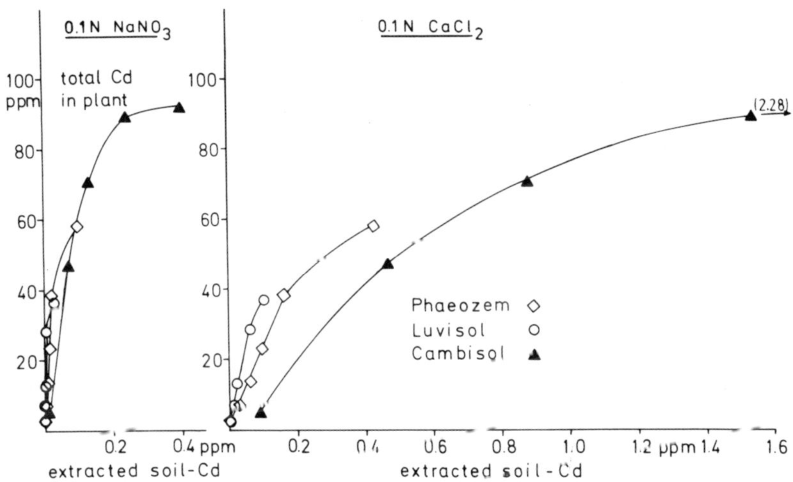

Figure VI: Cd-content of spinach versus soil-Cd extracted by $NaNO_3$ or $CaCl_2$

This was the reason for replacing the $NaNO_3$ by either 0.05 M or 0.1 M $CaCl_2$ in the German experiments (15). As the corresponding curves on Figures V and VI indicate, the amounts of Cd dissolved were considerably higher than with the $NaNO_3$, but still reflected the soil properties very well and also seemed to be proportional to the plant Cd uptake rates. Facing the fact that already 1 or 2 ppm total Cd in certain soils may be critical, the most important advantage of the $CaCl_2$ appears to be that it dissolves sufficiently large fractions even at these low contamination degrees in order to be easily measured without too much technical ado. Hence, the suitability of $CaCl_2$ for assessing the Cd availability from different soils and sources was tested in more detail, and the following promising results were obtained: Firstly, Figure VII demonstrates that an experimental variation of soil pH-values results in both differing plant uptake rates and amounts of soil Cd extracted at both $CaCl_2$ concentrations tested. The increases of uptake and solubility with increasing acidity are most obvious on the sandy cambisol, but even on the loamy luvisol the pH-effect was reproducibly reflected, especially if the more efficient 0.1 M $CaCl_2$ was used.

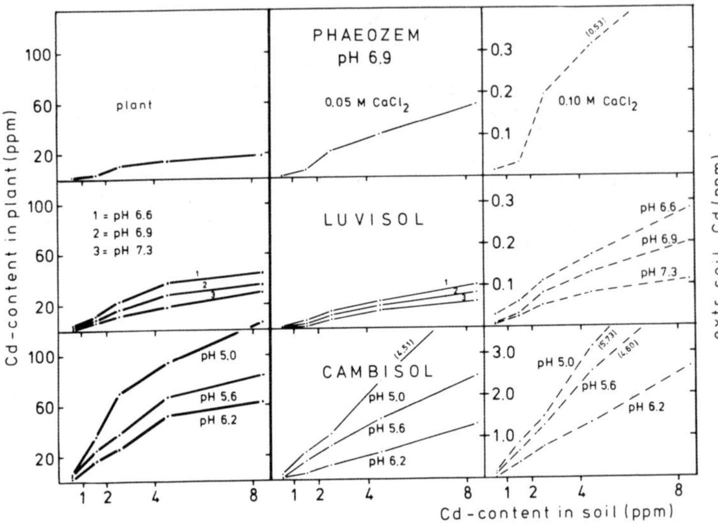

Figure VII: Influence of soil acidity on the Cd-contents of spinach and on $CaCl_2$-extractable soil-Cd at increasing Cd-levels (pot trial 1983)

Secondly, from Figure VIII it can be seen that the Cd availability from the artificially enriched sludge to ryegrass plants dropped from year to year during the course of the experiment, even though the pH of the soils in these pots decreased as well. This ageing of the Cd compounds (or perhaps better: their assimilation into soil-typical binding forms) was also reflected by the $CaCl_2$ extracts, although admittedly not as good as the pH effects in Figure VII.

Thirdly, the most different Cd uptake rates from the artificially enriched and the dredged sludge which was derived from contaminated sewage were also most clearly indicated by the $CaCl_2$ extractions according to Figure IX. The same is true for the almost identical uptake of Cd from the enriched sludge and from $CdSO_4$ on the neutral luvisol, whereas on the acid cambisol the uptake from $CdSO_4$ would be somewhat overestimated. Disregarding such minor shortcomings, however, there is no wonder that as a consequence of these beautiful results great expectations were raised that this $CaCl_2$ method might be the long-sought clue to give the desired realistic information about the actual Cd availability in different soils.

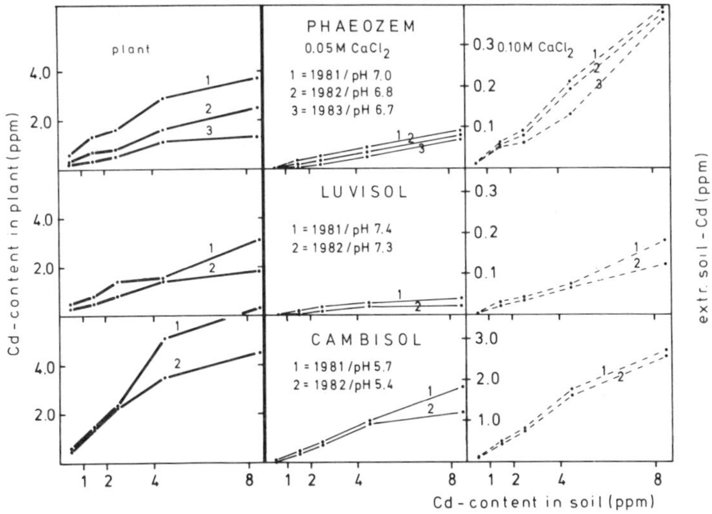

Figure VIII: Influence of reaction time on the Cd-contents of ryegrass (6. cut) and on $CaCl_2$-extractable soil-Cd at increasing Cd-levels (pot trial 1981 - 1983)

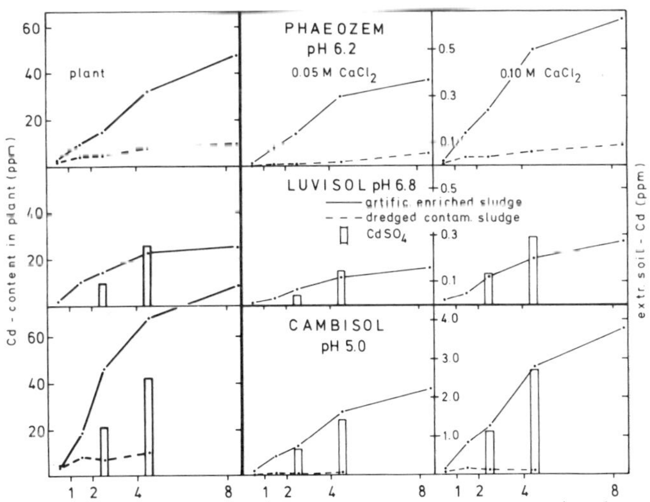

Figure IX: Influence of different Cd forms and origin on the Cd contents of radish leaves and on $CaCl_2$-extractable soil-Cd at increasing Cd-levels (pot trial 1981)

In fact, however, there is a limit to everything, although the plot on Figure X shows that the correlation between Cd uptake by spinach plants and $CaCl_2$-soluble soil Cd looked quite good even if the results from all three soils of the Braunschweig experiments were combined. The slightly curvilinear shape can be attributed to the fact that above 4 ppm Cd in these soils growth depressions started and the uptake of Cd might have been correspondingly depressed. Principally, however, such a close relationship, which includes not only the experimental Cd increments but also the additional treatments like pH variations and sludges of different origin, could hardly be expected to be any better in chemical soil testing research (15).

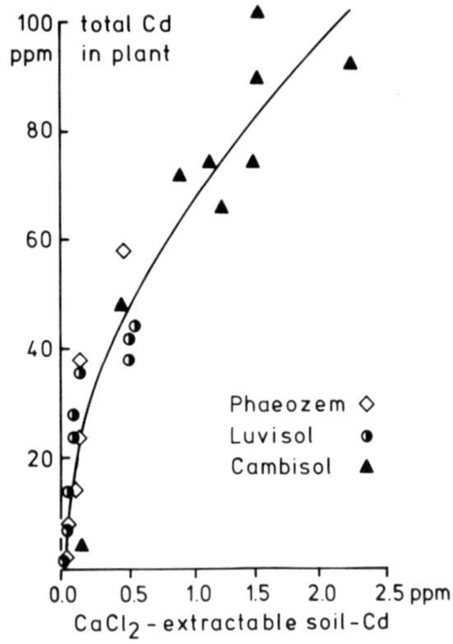

Figure X: Cd-content of spinach on 3 differently treated soils versus $CaCl_2$-extractable soil-Cd

Due to the many different treatments and variations the total number of observations from the Braunschweig pot experiment over a wide range of cadmium contents appeared to be large enough in order to justify a statistical treatment. The results of this calculation are given in Table V as r^2-values for the correlation between plant cadmium contents and total or $CaCl_2$-extractable soil-Cd, respectively. The experimental crops were arranged in the order of their Cd-uptake, starting from the accumulator plants spinach and radish or lettuce down to the much less sensitive grasses and to the corn salad which hardly translocated any Cd into its shoot tissue at all. As to the principle of the r^2-value it should be remembered that r^2-values of 0.5, 0.65, 0.80 and 0.90 correspond to correlation coefficients of 0.7, 0.8, 0.9 and 0.95, respectively.

It appears from these data that although only 3 soils were used, their differences in Cd availability were big enough in order to blur any relationship between total soil Cd contents and plant Cd uptake, which of course existed in the individual soils. On the other hand, both 0.05 and 0.1 M $CaCl_2$ reflected the Cd availability to most of the test plants with a reasonable degree of precision, except for the carrot roots and the grains of the wheat which are known to have relatively low heavy metal contents. The statistical results for lettuce and corn salad are, however, only of limited significance because these plants could not be grown on the acid cambisol.

Table V: r^2-values for the correlation between plant Cd-contents and total or $CaCl_2$-extractable soil-Cd
- pot experiments Braunschweig 1981 - 1983 -
(Overall results for 3 different soils, 5 - 8 Cd-increments in different forms and varied pH-adjustments)

crop plant and year	no. of values	total soil-Cd	0.05 M $CaCl_2$ extr.soil-Cd	0.1 M $CaCl_2$ extr.soil-Cd
spinach 1981 + 83	81	0.07	0.86	0.80
radish leaves 1981+83	85	0.01	0.72	0.83
" globes " "	85	0.03	0.83	0.81
lettuce 1982*	28	0.22	0.61	0.69
carrot leaves 1982	37	0.01	0.76	0.74
" roots "	37	0.02	0.37	0.61
spr.wheat str.1981+83	39	0.01	0.40	0.66
" " gr. " "	39	0.01	0.06	0.23
rye grass 6.cut 1981-83	46	0.18	0.65	0.75
corn salad 1982*	23	0.03	0.82	0.81

* = without cambisol (no growth)

This also happened to the other German participants in this joint pot experiment if soils below pH 5.5 were used. Regarding the other soils, however, the coordinated pot experiments at Bonn and Bremen were almost identical to the Braunschweig trial so that for an overall statistical evaluation all these experimental data could be combined. This has been done in Table VI with the result that here again no satisfactory correlation between total soil Cd and plant Cd uptake could be discerned. This correlation might have improved if a much larger number of observations were included as has been reported for instance for a joint investigation of working group A in the FAO network on trace elements (31).

Table VI: r^2-values for the correlation between plant and soil contents of cadmium if the data of Braunschweig, Bonn and Bremen are combined
- joint pot experiments 1981 - 1983 -
(overall evaluation of all available data)

crop plant	no. of values	total soil-Cd	0.1 M $CaCl_2$-extr. soil-Cd
spinach	95 (81)	0.31 (0.07)	0.79 (0.80)
radish leaves	181 (85)	0.16 (0.01)	0.64 (0.83)
" globes	154 (85)	0.21 (0.03)	0.43 (0.81)
lettuce	78 (28)	0.43 (0.22)	0.74 (0.69)
rye grass	135 (46)	0.23 (0.18)	0.61 (0.75)
corn salad	79 (23)	0.11 (0.03)	0.58 (0.81)

(figures in paranthesis = data from Braunschweig)

However, such a large-scale statistical approach still does not reflect the individual situations, whereas the limited number of $CaCl_2$-extractions on which Table VI is based certainly indicate a quite useful relationship. As in the previously described pot trials for Braunschweig the correlation between soil and plant Cd data was least for the plants or plant parts with low Cd incorporation, but the more Cd the plants absorbed, the better the correlation was. Thus there remains little doubt that the $CaCl_2$ extract gives a much more reliable information about the actual Cd availability in different agricultural soils than their total Cd contents do, even though a precise correlation for all possible cases cannot yet be expected.

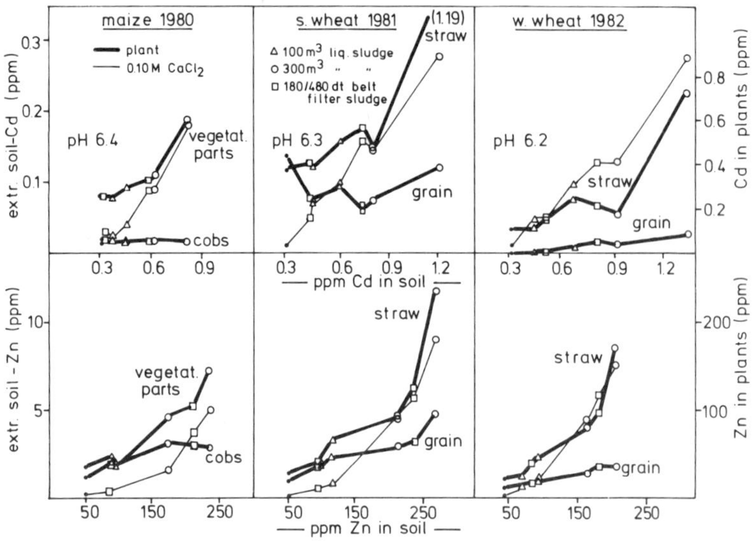

Figure XI: Plant contents and extractability of Cd and Zn on a cambisol field trial receiving increasing amounts of sewage sludge of different form and origin

The crucial problem, however, which questions all such model data from pot trials is of course whether they reflect the real conditions in the field. In order to test this we also collected and evaluated corresponding data from appropriate field experiments like the one at Braunschweig shown in Figure XI. Varying amounts of different sludges with low or high heavy metal contamination were used here year after year, so that increasing levels of metal loading have been obtained. Figure XI shows the relationship between total Cd and Zn contents in the soil and in 3 crop plants, together with the amounts of Cd and Zn which were released by 0.1 M $CaCl_2$. Just as in other reports (35, 36, 37, 38) the heavy metal accumulation in grains was quite low, but the Cd and Zn contents of the vegetative plant parts increased significantly with the corresponding soil contents. Some minor deviations from this overall trend were due to differing pH values as a consequence of differing sludge materials, but the general relationship between plant uptake and solubility in $CaCl_2$ looked quite clear.

This happened not only in this particular but also in another similarly designed field experiment (39), so that according to Table VII the

$CaCl_2$ extracts gave much better correlations with the actual plant uptake than the total soil contents did for both Cd and Zn not only in pots but also in the field. A statistical evaluation with such small numbers of observation pairs on 2 similar soils has of course its severe limitations, but it nontheless indicates that where Zn becomes a problem, a $CaCl_2$ extraction may be at least as informative for this as for the element Cd and certainly much more dependable than data for total soil contents only.

Table VII: r^2-values for the correlation between plant and soil contents of Cd and Zn (total and $CaCl_2$-extractable)
- sewage sludge field experiments at Braunschweig on 2 different soils, receiving increasing amounts of sludge of different forms and origin

crop plant and year	no. of values	Cadmium total	Cadmium $CaCl_2$-ext.	Zinc total	Zinc $CaCl_2$-ext.
maize 1980 veg.parts	12	0.26	0.92	0.35	0.97
" " cobs		0.04	0.58	0.56	0.53
spr.wheat 1981 straw	12	0.28	0.43	0.38	0.88
" " " grain		0.11	0.09	0.53	0.64
w.wheat 1982 straw	12	0.43	0.85	0.41	0.94
" " " grain		0.70	0.81	0.70	0.81
maize 1983 veg.parts	12	0.14	0.77	0.31	0.85
" " cobs		0.02	0.27	0.41	0.77

In order to test this on a broader base it would be most desirable to compile and compare more such results from similar field experiments with heavy metals as has been tried by the Subgroup B of the FAO Trace Element Network for its consultation at Aarhus in 1983 (34, 40). Background data for some such comparable German sludge experiments with reasonably increased heavy metal contents are listed in Table VIII. Hoping to widen the spectrum of samples a number of additional data pairs from field pollution surveys has also been testwise included, although without complete information about all sampling details.

Table VIII: Some long-term field experiments with sewage sludge or town waste compost in the Fed. Rep. Germany, and field pollution surveys rendering information on heavy metals loading of soils and plants

location		no. of treatm.	pH	% C	CEC	material used
Braunschweig						liquid sludge
	Bs1	7	5.5	1.6		belt filter sludge
	Bs2	5	6.7	0.9		liquid sludge
Bremen	Br1	2	4.2	28.9	234	liquid sludge
	Br2	3	4.0	16.1	190	" "
Gießen	Gi1	3	6.6	1.1	11.6	sludge-garbage comp.
	Gi2	3	6.2	0.9	11.6	dewatered sludge
	Gi3	3	5.9	0.8	11.6	liquid sludge
	Gi4	3	5.4	5.9	11.6	" "
München	Mü1	6	5.8	1.7	11	centrif. sludge
	Mü2	4	6.6			filter sludge
Field		19	contaminated by river floods			
pollution	Su	11	naturally elevated contents			
survey		14	immission-caused contamination			

The trouble with all such field experiments is that they have been designed and carried out in most different ways, that their rates of heavy metal loading are very different, and that only a few comparable crops have been grown on them. Table IX lists a number of soils and observation pairs which could be compared, and the ranges of total Cd and Zn contents obtained. The abbreviations refer to those in Table 8. Although the ranges of soil contamination and plant contents look rather broad, the distribution of individual values along this range is not uniform at all. For the time being, however, this is the best set of data available, and hence a statistical treatment in order to test the correlation between soil and plant Cd or Zn contents was tried.

Table IX: Information available from sewage sludge experiments and field pollution surveys in the Fed. Rep. Germany for assessing the relationship between Cd- and Zn-contents of soils and plants

material	sources	number of soils	values	range of soil-Cd	range of plant-Cd	range of soil-Zn	range of plant-Zn
lettuce	Mü1, Su	5	17	0.7-23	0.4-14.5	15-805	64-204
maize stems	Bs 1 + 2	2	24	0.2-1.7	0.1-1.2	34-240	13-314
" cobs		2	24		0.05-0.25		28-137
s.wheat straw	Bs1+2, Mü1+2,	5	27	0.2-50	0.4-5.8	14-1695	18-292
" grain	Su	5	27		0.1-3.1		29-145
carrot roots	Su	8	13	1.2-6.3	0.1-1.3	253-1230	23-61
sug.beet roots	Gi3, Mü1, Su	4	17	0.1-23	0.1-2.6	30-900	23-148
leek	Su	8	13	1.2-6.3	0.2-1.1	253-1230	29-276
w.wheat straw	Bs1+2, Mü2,	6	19	0.2-50	0.1-3.5	35-1695	12-336
" grain	Su	6	19		0.01-1.9		12-125
oats straw	Gi1+2, Mü1+2	4	16	0.1-50	0.04-3.5	30-1695	21-245
" grain		4	21		0.02-2.7		32-121
kale	Su	9	12	1.2-6.3	0.1-0.9	332-1230	46-181
pasture 1.cut	Br1+2, Gi4,	12	23	0.5-50	0.10-1.8	38-1695	40-191
grass 2.cut	Mü2, Su	10	18			94-1695	33-219
s.barley straw	Mü1+2, Su	4	12	0.7-50	0.1-2.2	30-1695	23-272
" grain		4	18		0.02-0.5		31-128
potato tubers	Mü1+2	2	15	0.7-50	0.3-1.0	15-1695	14-46

Table X summarizes the results and confirms once again that in most cases plant Cd or Zn levels are poorly correlated with total soil contents of these 2 elements. The correlation becomes much better for the $CaCl_2$ soluble fraction of both Cd and Zn in soil at least for those plants or plant parts which absorb and contain significant amounts, whereas for roots and grains the Zn results were less encouraging. There are of course many limitations in this array of data especially as far as the accidentally chosen survey samples are concerned. However, any one who is familiar with such practical limitations will agree that this nevertheless proved to be an encouraging approach which is worthwhile to follow in more detail.

In conclusion it may be stated that dilute $CaCl_2$ is a most promising solvent in order to reflect the plant availability of Cd from very different soils under both pot and field conditions, and that, although with somewhat less precision, this may also apply for Zn. The acquisition and comparison of appropriate field data is, however, difficult, and therefore an international cooperation to widen this data base is urgently needed.

Table X: r^2-values for the correlation between plant and soil contents of Cd and Zn (total and $CaCl_2$-extractable) if the available field and survey data are combined

material	sources	cadmium total	cadmium $CaCl_2$ ext.	zinc total	zinc $CaCl_2$ ext.
lettuce	Mü, Su	0.03	0.51	0.43	0.53
maize stems	Bs	0.26	0.92	0.32	0.64
" cobs		0.01	0.58	0.37	0.72
s.wheat straw	Bs, Mü, Su	0.42	0.53	0.05	0.67
" grain		0.46	0.52	0.46	0.15
carrot roots	Su	0.15	0.74	0.02	0.03
sug.beet roots	Gi, Mü, Su	0.07	0.62	0.01	0.14
leek	Su	0.21	0.72	0.01	0.01
w.wheat straw	Bs, Mü, Su	0.24	0.76	0.02	0.56
" grain		0.59	0.41	0.70	0.23
oats straw	Gi, Mü	0.79	0.34	0.85	0.34
" grain		0.59	0.41	0.70	0.23
kale	Su	0.12	0.49	0.01	0.17
pasture 1.cut	Br, Gi,	0.74	0.32	0.10	0.04
grass 2.cut	Mü, Su	0.58	0.35	0.14	0.21
s.barley straw	Mü, Su	0.11	0.14	0.81	0.24
" grain		0.21	0.29	0.85	0.11
potato tubers	Mü	0.03	0.09	0.55	0.09

These investigations have been carried out as part of the EC coordinated program COST 68 (bis) and were financed by the German Federal Office of the Environment (Umweltbundesamt).

REFERENCES

1. KLOKE, A., SAUERBECK, D. R. and VETTER, H. (1984). The contamination of plants and soils with heavy metals and the transport of metals in terrestrial food chains. In: Nriagu, J. O. (Ed.): Changing metal cycles and human health. Berlin, Springer Verlag, pp. 113 - 141.
2. DAVIS, R. D., HUCKER, G. and L'HERMITE, P. (1983). Environmental effects of organic and inorganic contaminants in sewage sludge. EUR 8022, D. Reidel Publishing Co.
3. WEBBER, M. D., KLOKE, A. and TJELL, J. Chr. (1984). A review of current sludge use guidelines for the control of heavy metal contamination in soils. Proc. 3rd Intern. Symposium "Processing and Use of Sewage Sludge", Brighton/UK, 27 - 29 September 1983 (in press).
4. HÄNI, H. and GUPTA, S. (1983). Second report on the standardized Cd-pot experiment. EEC Concerted Action (COST), Working Group 5, February 1983.
5. HERMS, H. and BRÜMMER, G. (1980). Einfluß der Bodenreaktion auf Löslichkeit und tolerierbare Gesamtgehalte an Nickel, Kupfer, Zink, Cadmium und Blei in Böden und kompostierten Siedlungsabfällen. Landwirtsch. Forsch. 33, 408 - 423.
6. STYPEREK, P., SAUERBECK, D. and TIMMERMANN, F. (1983). Cd-Verfügbarkeit in verschieden behandelten Böden in Abhängigkeit von Menge und Bindungsform. Landwirtsch. Forsch. 39, 183 - 195.

7. HÄNI, H. and GUPTA, S. (1980). Ein Vergleich verschiedener methodischer Ansätze zur Bestimmung mobiler Schwermetallfraktionen im Boden. Landwirtsch. Forsch. Sh. 37, 267 - 274.
8. COTTENIE, A. and KIEKENS, L. (1981). Beweglichkeit von Schwermetallen in mit Schlamm angereicherten Böden. Korrespondenz Abwasser 28, 206 - 210.
9. KIEKENS, L. and COTTENIE, A. (1982). Principles of soil analysis with regard to trace elements. In: Newsletter from the FAO European Cooperative Network on Trace Elements. State University Gent/Belgium.
10. LAKANEN, E. and ERVIOE, R. (1971). A comparison of eight extractants for the determination of plant available micronutrients in soils. Acta Agr. Fenn. 123, 223 - 232.
11. COTTENIE, A. (1978). Soil and plant testing as a basis of fertilizer recommendation. FAO, Rome, p. 90.
12. SAUERBECK, D. and RIETZ, E. (1983). Soil-chemical evaluation of different extractants for heavy metals in soils. In: R. D. Davis, G. Hucker and P. L'Hermite (eds.) Environmental effects of organic and inorganic contaminants in sewage sludge. Proc. of a workshop held at Stevenage, May 25 - 26, 1982. Dordrecht, Boston, London: D. Reidel Publ. Comp. pp. 147 - 160.
13. GUPTA, S., HÄNI, H. and SCHINDLER, P. (1980). Mobilisierung und Immobilisierung von Metallen durch die organische Bodensubstanz. Eidg. Forschungsanstalt für Agrikulturchemie und Umwelthygiene, Liebefeld-Bern, pp. 32 - 47.
14. KÖSTER, W. and MERKEL, D. (1983). Beziehungen zwischen den Gehalten an Zink, Cadmium, Blei und Kupfer in Böden und Pflanzen bei Anwendung unterschiedlicher Bodenuntersuchungsmethoden. Landwirtsch. Forsch. Sh. 39, 245 - 254.
15. SAUERBECK, D. R. and STYPEREK, P. (1984). Predicting the cadmium availability from different soils by $CaCl_2$-extraction. Proc. 3rd Intern. Symposium "Processing and Use of Sewage Sludge", Brighton/UK, 27 - 29 September 1983 (in press).
16. DAVIS, R. D. and COKER, E. G. (1980). Cadmium in agriculture with special reference to the utilization of sewage sludge on land. Stevenage Laboratory, Water Research Centre, Technical Report TR 139, Stevenage/UK.
17. KLEIN, H., PRIEBE, A. and JÄGER, H.-J. (1981). Grenzen der Belastbarkeit von Kulturpflanzen mit dem Schwermetall Cadmium. Angew. Botanik 55, 295 - 308.
18. KLOKE, A. and SCHENKE, H.-D. (1979). Der Einfluß von Cadmium im Boden auf den Ertrag verschiedener Pflanzenarten und deren Cadmiumgehalt. Z. Pflanzenernähr. Bodenkde. 142, 131 - 136.
19. MACLEAN, A. J. (1976). Cadmium in different plant species and availability in soils as influenced by organic matter and addition of lime. Canad. J. Soil Sci. 56, 129 - 138.
20. SAUERBECK, D. (1983). Welche Schwermetallgehalte in Pflanzen dürfen nicht überschritten werden, um Wachstumsbeeinträchtigungen zu vermeiden? Landwirtsch. Forsch. Sh. 39, 108 - 129.
21. BAKER, A. J. M. (1981). Accumulators and excluders - strategies in the response of plants to heavy metals. J. Plant Nutrition 3, 643 - 654.
22. BINGHAM, F. T., PAGE, A. L., MAHLER, R. J. and GANJE, T. J. (1975). Growth and cadmium accumulation of plants grown on a soil treated with a cadmium-enriched sewage sludge. J. Env. Qual. 4, 207 - 211.
23. COTTENIE, A. and CAMERLYNCK, R. (1981). Specific behaviour of some monocotyledon and dicotyledon species with regard to excess of trace elements. In: A. Cottenie (ed.) Trace Elements in Agriculture and in the Environment. Gent/Belgium: Laboratory of Analytical and Agrochemistry, pp. 43 - 50.

24. DAVIS, R. D. and CARLTON-SMITH, C. (1980). Crops as indicators of the significance of contamination of soil by heavy metals. Stevenage Laboratory, Water Research Centre, Technical Report TR 140, Stevenage/UK.
25. OBERLÄNDER, H.-E. and ROTH, K. (1982). Uptake and distribution of soil-applied labelled heavy metals in cereal plants and products. FAO/IAEA Research Coordination Meeting "Agrochemical Biota Interactions in Soil and Water Using Nuclear Techniques", Technical Document No. 263, pp. 137 - 149. Vienna: International Atomic Energy Agency.
26. OCKER, H.-D., SEIBEL, W. and BRÜGGEMANN, J. (1983). Reduktion des Schwermetallgehaltes durch die Getreideverarbeitung. Landwirtsch. Forsch. 39, 333 - 341.
27. KIEKENS, L. (1979). Adsorption, desorption and distribution of trace elements in soils in function of some soil parameters. In: A. Cottenie et al.: Essential and non essential trace elements in the system soil - water - plant. Lab. of Analytical and Agrochemistry, Gent/Belgium, pp. 23 - 43.
28. KLOKE, A. (1981). Aufnahme umweltrelevanter Elemente durch die Pflanze. Deutsche Gesellschaft für Qualitätsforschung (Pflanzliche Nahrungsmittel) e. V., XVII. Vortragstagung "Siedlungsabfall-Verwertung und Nahrungsqualität", Speyer, 26. - 27.03.1981, pp. 65 - 90.
29. PLUQUET, E. (1984). Die Bedeutung des Tongehaltes und des pH-Wertes für die Schwermetallaufnahme einiger Kulturpflanzen aus kontaminierten Böden. Umweltforschungsplan des Bundesministeriums des Innern. Abfallwirtschaft-Forschungsbericht 103 01 210, UBA-FB-82-035.
30. COTTENIE, A., KIEKENS, L. and van LANDSCHOOT, G. (1984). Problems of the mobility and predictability of heavy metal uptake by plants. Proc. 3rd Intern. Symposium "Processing and Use of Sewage Sludge", Brighton/UK, 27 - 29 September 1983 (in press).
31. KIEKENS, L. and COTTENIE, A. (1984). Principles of investigations on the mobility and plant uptake of heavy metals. EC-Seminar "Chemical Methods for Assessing Bio-available Metals in Sludges and Soils", Münster, 11 - 13 April (in press).
32. HÄNI, H. and GUPTA, S. (1984). Choice of an extractant for simulating the availability and absorption of heavy metals by plants. Proc. 3rd Intern. Symposium "Processing and Use of Sewage Sludge", Brighton/UK, 27 - 29 September 1983 (in press).
33. HÄNI, H. and GUPTA, S. (1984). Reasons to use neutral salt solutions to assess the metal impact on plants and soils. EC-Seminar "Chemical Methods for Assessing Bio-available Metals in Sludges and Soils", Münster, 11 - 13 April (in press).
34. Report of the 1983 Consultation on the European Cooperative Network on Trace Elements held in Aarhus, Denmark, 31 May - 3 June. Food and Agriculture Organization of the United Nations.
35. DIEZ, Th. (1982). Einfluß der Klärschlammbehandlung und der Beschlammungsintensität auf den Schwermetalltransfer Boden - Pflanze. Landwirtsch. Forsch. Sh. 39, 213 - 223.
36. HERMS, U. and TENT, L. (1982). Cadmium-Gehalte in Spülfeldern aus Hafenschlick und in darauf angebauten Kulturpflanzen - eine Felderhebungsuntersuchung. Landwirtsch. Forsch. Sh. 39, 448 - 456.
37. KRÄMER, F. and KÖNIG, W. (1982). Cadmiumgehalte in Böden und Pflanzen auf klärschlammgedüngten landwirtschaftlichen Nutzflächen. Landwirtsch. Forsch. Sh. 39, 434 - 447.
38. SCHAAF, H. and v. BOGUSLAWSKI, E. (1983). Schwermetallanreicherung in Boden und Pflanzen bei langjähriger Anwendung von Klärschlamm. Landwirtsch. Forsch. Sh. 39, 224 - 237.

39. SÖCHTIG, H. and TIMMERMANN, F. (1984). Schwermetalleintrag in Böden und -aufnahme durch Pflanzen bei landwirtschaftlicher Klärschlammverwertung. Angewandte Botanik (in press).

40. SAUERBECK, D. and DIEZ, T. (1984). Heavy metal contents of soils and plants from field experiments receiving large amounts of sludge or town waste composts in the FRG. In: Newsletter from the FAO European Cooperative Network on Trace Elements. State University Gent/Belgium (in press).

ACKNOWLEDGEMENT

Our heartfelt thanks are extended to Prof. Dr. H. Kuntze, Bremen, Prof. Dr. Dr. h. c. E. von Boguslawski, Rauisch-Holzhausen, Dr. Th. Diez, Munich, and Mr. W. König, Düsseldorf, for their generous supply of data and soil samples from their field experiments and pollution surveys.

DISCUSSION

Chairman: R. LESCHBER

Rapporteurs: G.C. PORTEOUS, T.M. ADAMS, J.R. SAUNDERS

It was observed that a particularly good correlation was found between extractability of Cd by calcium chloride and plant concentrations of Cd on polluted soils but the correlation for uncontaminated soils was less satisfactory. One explanation for this was that bioavailable Cd in polluted soils was probably held by exchange processes and could be displaced by Ca^{2+}. Another consideration was the formation of the $CdCl^+$ complex with its high stability constant which might mean that calcium chloride was really suitable only as an extractant for Cd. Although it was thought possible that chlorocomplexes were formed the choice of calcium chloride as opposed to other calcium salts had been largely empirical. Calcium nitrate had been tried but higher concentrations of this salt were needed. The important factor was the activity of divalent Ca. Use of calcium chloride showed promise for Zn but preliminary results for Ni were not very satisfactory.

Discussion of the influence of soil properties suggested that apart from pH value the presence of oxides of iron and manganese also affected the bioavailability of Cd. It seemed that iron oxides were particularly important in reducing the bioavailability of Cd and when heavy dressings of sludge were used to reclaim low-Fe sandy soils, the addition of Fe compounds as well as lime should be considered to minimise metal availability. It was thought that air-drying of soils before extraction could affect the influence of Fe and Mn oxides and special procedures, i.e. extracting moist soil might need to be considered to investigate their effects.

Correlations between Cd concentrations in plants and extractability of Cd by calcium chloride were less satisfactory in field conditions than in pot experiments and possible reasons for this were discussed. A major factor was thought to be loss of control over plant growing conditions in the field and the fact that the soil explored by plant roots in field conditions was less easily defined than the homogenised contents of a glasshouse pot. It was agreed that further evaluation of calcium chloride in field conditions would be desirable and delegates were requested to supply Professor Sauerbeck with relevant data from their field trials if possible. So far, evaluation of the method had concentrated on Cd but it was expected to perform satisfactorily for Zn also and perhaps Tl. Pb and Cr were less important in this context because their bioavailability in sludge-treated soil was minimal. There was interest in repeated extractions with calcium chloride which might extend the usefulness of the technique by giving an indication of the long-term bioavailability of Cd as well as its immediate bioavailability. It was noted that calcium chloride extracted more Cd from the soil than did sodium nitrate. Therefore calcium chloride might be the better of these two neutral salt extractants for assessing the bioavailability of Cd in background and slightly contaminated soils because the analytical problems (except for the presence of high concentrations of Ca) would not be so great.

(Contributors to discussion : Tjell, Sanders, Kick, Gomez, Kuntze, Hani, Webber, Fleming, Worthington, Tarradellas, Hardwick, Salt).

THE EFFECT OF INCUBATION ON THE COMPOSITION OF SOIL SOLUTION DISPLACED FROM 4 SOILS TREATED WITH ZINC, COPPER OR NICKEL-LOADED SEWAGE SLUDGE

T. McM. Adams and J.R. Sanders
Rothamsted Experimental Station, Harpenden, Hertfordshire, AL5 2JQ, U.K.

Summary

Soil solution was displaced from a calcareous silt, a clay loam and 2 sandy loam soils at 1, 3, 7, 11, 16 and 21 months after metal-loaded sewage sludges had been applied. The soluble organic C contents decreased markedly over the first 3 months, then stabilised, and were similar in all soils after 11 months. The pH values of soil solution from all sludge-treated soils decreased to some extent over 7 months, thereafter values from 2 soils stabilised about pH 7.5 and the other 2 decreased to about 6.0. The soil solution concentration of zinc, copper and nickel was affected by the total metal added to the soil, by the pH of the soil solution, and by soil properties, such as texture, mineralogy, and content of organic matter and free oxides of Fe, Al and Mn. The pH values of soil cores taken from replicate pots and their DTPA-extractable metal values did not generally show such consistent trends with time as the soil solution concentrations did.

When soil solution concentrations, DTPA-extractable metal, and total metal content of soil were compared as indices of metal availability to clover plants, all 3 methods were equally good for a single soil but when data from all 4 soils were used soil solution was the best index, e.g. for zinc in all 4 soils r = 0.92 for soil solution, 0.54 for DTPA-extractable, and 0.41 for total metal contents.

1. INTRODUCTION

The present study uses displaced soil solution to examine the release of metal from soils treated with zinc, copper or nickel-loaded sewage sludge as the period of incubation increases. This forms part of a wider project on the availability to plants of metals added to soil in sludge, and of the soil properties influencing such availability.

Examination of the soil solution is an attractive technique as it does not involve the use of strong chemicals or empirical reagents such as EDTA or DTPA, and is especially suited to examination of contaminated soils where analyses are easier. Furthermore, elements in the soil solution are believed to be immediately available to plants or in rapid equilibrium with forms which are taken up. The use of the displacement method for obtaining solution permits non-destructive sampling of the same soil samples at intervals over a long period, and processing of a large number of samples on each occasion. Guidelines already exist for the amount of metal which may be added to soil in sewage sludge (7), but changes with time in the fraction of the metals in solution could indicate long term hazards, and it is this aspect which is studied here.

2. METHODS

Four sludges were prepared from similar batches of raw sewage, namely a control sludge, and a copper, nickel, and zinc-loaded sludge (2). The soils used were two sandy loams (Cottenham and Newport series), a clay loam (Carswell series) and a calcareous silt loam (Upton I series). The metal-loaded sludges were either added directly to the soils to give the high-metal treatments or after mixing 1:1 with control sludge to give the low-metal treatments. The zinc-loaded sludge was mixed with either the copper-loaded or nickel-loaded sludge to give a combined low zinc-low copper or low zinc-low nickel treatment. Control-sludge and no-sludge treatments were also included. The sludges were thoroughly mixed with the soils at a sludge dry solids:soil ratio of 1:100. The sludge-treated soils were then air-dried, re-sieved and 1 kg lots were weighed into plastic pots. Four fallow replicates were kept under black plastic sheeting in a glasshouse at a temperature of approximately 20°C, 2 being used for soil solution displacement and 2 for provision of soil cores. The water holding capacity was adjusted to, and kept at, 90% by addition of deionised water.

The soil solution was displaced after 1, 3, 7, 11, 16 and 21 months by gradual addition of small aliquots of deionised water to the top of the soil, to ensure that the water infiltration rate was slow and that the water entered the fine capillaries. This is necessary to achieve the "piston effect" when displacing the soil solution. The first 10 mls of solution were discarded, and a further 40 mls were collected and centrifuged at 5000 g for 10 minutes. Solution was displaced from fallow replicates and also, less frequently, from pots where 1 barley and 3 red beet crops had been grown, and once, from others where clover had been grown.

The pH was measured by glass electrode, and the soluble organic carbon by dichromate oxidation. The solutions were also analysed for P, K, Ca, Mg, Na, S, Zn, Cu and Ni using inductively coupled plasma (ICP) spectrometry, and also for Cl and NO_3 using a "Technicon" auto-analyser. The major element concentrations were required to calculate ionic strength for speciation of zinc, copper and nickel by the ion-exchange technique (6). Nickel was determined in some solutions by graphite furnace atomic absorption spectrometry. The soils were analysed for total metal content by ICP spectrometry after aqua regia digestion (Table I) and also extracted with 0.005M DTPA at pH 7.3 at a soil:solution ration of 1:2 (4).

Table I. Total metal contents of soils in mg metal kg^{-1} soil.

Soil	Sludge	Zn	Cu	Ni	Al	Mn	Fe
Cottenham	None	55	13	13	5551	234	17465
	Control	72	18	14	5571	238	19503
	Low zinc	214	18	13	5621	222	17811
	High zinc	397	22	13	5371	221	16683
	Low copper	73	79	14	5406	240	19881
	High copper	65	96	15	5573	231	20021
	Low nickel	68	16	43	5709	228	17672
	High nickel	64	15	77	6024	257	21262
Newport	None	34	6	8	5514	270	9374
	Control	51	11	9	5872	270	9245
	Low zinc	183	12	6	6057	281	9441
	High zinc	375	21	7	6007	278	9806
	Low copper	51	75	6	6027	286	9460
	High copper	49	143	5	5834	271	9348
	Low nickel	45	9	33	6248	283	9537
	High nickel	42	8	66	6595	286	9515
Carswell	None	152	29	49	38024	396	43909
	Control	162	33	49	37983	427	43164
	Low zinc	400	39	47	39162	333	44400
	High zinc	483	41	48	38078	361	44904
	Low copper	164	93	49	39129	426	44283
	High copper	165	179	50	38546	463	46093
	Low nickel	161	32	75	39072	469	44679
	High nickel	161	32	109	30437	555	46727
Upton	None	81	14	13	9068	446	9284
	Control	95	18	14	9377	456	9316
	Low zinc	248	22	13	9248	458	9233
	High zinc	381	25	13	9186	457	9487
	Low copper	97	85	14	9297	454	9262
	High copper	97	173	13	9388	456	9101
	Low nickel	94	18	45	9682	458	9152
	High nickel	88	15	69	9546	447	8908

3. RESULTS AND DISCUSSION

3.1. pH and soluble organic C contents of solution from fallow treatments

The pH values of untreated replicates did not show any trends with time, except possibly in the clay loam soil (Carswell, Table II). However, in the sludge-treated soils the pH decreased as the period of incubation increased up to 7 months in all soils (Tables II and III). After this period the pH of Cottenham and Carswell soil solutions decreased further to 6.0 and 6.2 respectively, but the pH of Newport and Upton soil solutions stabilised at about 7.4 and 7.5 respectively (Table III).

The soluble organic C contents in both untreated and treated soils decreased over the first 3 months. In the untreated soils this was due to the effects of air-drying and re-wetting causing a flush of decomposition of soil organic matter. Initial organic C levels in soil solutions from sludge treatments were much higher than in untreated replicates, but they decreased sharply with time, and all soils eventually stabilised at similar values after 11 months. The decomposition of the sludge organic matter may be the cause of the sharp fall in pH in sludge-treated replicates, by production of sulphate, nitrate, and CO_2 during aerobic decomposition. The Upton and Newport soils were better buffered than the Cottenham and Carswell soils, which would be expected for the calcareous Upton soil. The difference in behaviour between the two light-textured soils (Cottenham and Newport) is difficult to explain, though the Newport soil had a higher total organic matter content, which may provide better buffering.

3.2. Metal concentrations in soil solution from fallow treatments

The 3 factors controlling the concentration of metal in solution were amount of metal added to soil, pH and soil type, which is regarded here as including texture, mineralogy, and content of organic matter and free oxides of iron, aluminium and manganese. These factors may affect particular metals differently.

a) The effect of addition of metal-loaded sewage sludge

The addition of sludge organic matter alone did not affect the metal concentrations in soil solutions, as these were similar in solutions from untreated and control sludge-treated soils, except where pH values were markedly different (Table II). The addition of metal-enriched sludges gave greatly increased concentrations of metals in the soil solution (Table III). In general, the addition of a high-metal sludge to a soil gave larger increases in solution concentration than did addition of the corresponding low-metal sludge, though comparison was complicated by differences in pH values.

b) Comparison of concentrations of zinc, copper and nickel

The relative concentrations of zinc and nickel in soil solution reflected the quantities of metal added in sludge. However, the copper concentrations were generally lower than those of nickel even though the quantity of copper added to the soil was three times that of nickel.

Table II. Composition of soil solution displaced from untreated replicates, and those treated with control sludge.

Period since sludge application	No treatment					Control sludge treatment				
	pH	Soluble C	Zn	Cu	Ni	pH	Soluble C	Zn	Cu	Ni
Months		mg C l^{-1}	μM	μM	μM		mg C l^{-1}	μM	μM	μM
			COTTENHAM							
1	6.5	28	32.1	0.15	<0.26	7.6	72	6.2	0.07	0.28
3	5.7	11	10.3	0.05	<0.26	7.4	31	7.4	0.15	<0.26
7	6.2	15	6.3	0.13	0.46	6.8	23	7.5	0.21	0.81
11	6.0	13	4.9	0.3	0.03	6.5	25	6.2	0.41	0.35
16	6.2	13	3.2	0.4	0	6.2	24	6.1	0.31	0
21	6.0	11	2.4	0.37	0.84	6.4	24	4.5	0.47	0.53
			NEWPORT							
1	7.7	48	9.0	0.12	<0.26	7.9	46	17	0.04	<0.26
3	7.5	22	2.1	0.05	<0.26	7.9	29	3.3	0.22	<0.26
7	7.6	20	1.6	0.09	0.4	7.6	18	2.6	0.23	0.99
11	7.3	18	0.5	0.23	0.02	7.3	16	1.5	0.37	0.04
16	7.5	17	0.7	0 13	0	7.4	20	1.6	0.22	0
21	7.5	17	0.6	0.49	0.97	7.0	16	1.7	0.55	0.94
			CARSWELL							
1	7.2	-	14.0	0.12	<0.26	7.5	-	11	0.13	<0.26
3	6.7	16	4.4	0.02	<0.26	7.1	21	4.2	0.03	<0.26
7	6.5	12	6.6	0.05	0.58	7.2	17	3.0	0.06	0.68
11	6.7	10	3.4	0.31	0.35	6.8	19	2.1	0.31	0.16
16	6.6	13	1.8	0.33	0.45	6.7	21	1.8	0.21	0.28
21	5.9	7	2.7	0.1	0.46	6.3	16	2.5	0.14	0.52
			UPTON							
1	7.9	35	10.3	0.08	<0.26	7.9	19	4.6	0.13	<0.26
3	7.8	16	2.2	0.02	<0.26	7.9	27	2.2	0.06	<0.26
7	7.8	13	1.2	0.05	0.39	7.7	14	1.3	0.1	0.64
11	7.5	14	0.9	0.21	0.04	7.3	17	0.6	0.34	0.15
16	7.8	18	0.4	0.14	0.37	7.7	16	0.7	0.14	0.23
21	7.7	14	0.5	0.19	0.43	7.5	16	0.9	0.24	0.54

Table III. Composition of soil solution displaced from fallow treatments at various times after application of metal-loaded sewage sludge

Period since sludge application	pH	Soluble C	Sludge treatment					
			Zinc		Copper		Nickel	
			Low	High	Low	High	Low	High
Months		mg C l^{-1}	μM	μM	μM	μM	μM	μM
			COTTENHAM					
1	7.4	72	5.1	29	0.38	2.9	3.9	5.7
3	7.2	33	8.6	10	0.97	1.3	1.6	3.8
7	6.7	19	50	54	0.50	1.6	3.7	8.2
11	6.7	20	48	51	0.9	2.1	4.9	8.0
16	6.2	21	72	71	0.8	2.0	7.6	12.3
21	6.0	17	56	65	0.75	2.0	6.5	10.4
			NEWPORT					
1	7.8	83	11	17	0.3	2.0	2.0	3.4
3	8.0	39	3.2	6	0.9	2.2	0.8	1.6
7	7.6	25	6.3	10	1.3	2.2	1.9	2.3
11	7.3	20	2.7	7	1.7	1.7	1.3	1.6
16	7.4	18	3.8	13	1.0	1.3	1.0	1.7
21	7.3	18	3.1	10	0.9	1.5	1.6	1.6
			CARSWELL					
1	7.6	-	12	19	0.5	0.6	1.5	1.8
3	6.6	20	6	5.3	0.6	0.7	<0.26	0.9
7	7.1	18	7	7	0.4	0.6	1.3	1.9
11	6.8	16	5.7	6.7	0.6	0.8	0.8	1.3
16	6.6	20	6.9	6.9	0.6	0.8	1.2	1.9
21	6.2	12	9.5	16.8	0.4	0.7	1.6	2.6
			UPTON					
1	8.0	37	7.3	8.7	0.3	1.1	0.68	1.8
3	7.8	21	2.8	3.0	0.8	1.2	0.53	1.0
7	7.9	15	2.4	2.1	0.4	0.6	1.4	1.6
11	7.4	14	1.6	1.9	0.7	0.8	0.6	0.9
16	7.6	15	1.2	1.6	0.4	1.0	0.6	0.85
21	7.5	14	1.2	1.8	0.4	0.7	0.9	1.4

This suggests that the strengths of retention are very different for copper on the one hand and zinc and nickel on the other. Adams & Sanders (1) found that the mechanisms of retention were different for copper. McBride and Blasiak (5) concluded that zinc solubility in soil was mainly controlled by specific adsorption on hydrous oxides of iron and aluminium below pH 6.5, but that copper concentrations were controlled by adsorption on insoluble organic matter.

c) The effect of soil type

The effects of soil type can be seen at the 1 month sampling when addition of identical amounts of metal-loaded sludge to soils at similar pH values resulted in a range of metal concentrations (Table III). The differences between the 2 light-textured soils (Cottenham and Newport) were due to differences in organic matter and free oxide contents (3). Further, in Cottenham and Carswell soils the variation of pH with time was similar but the concentrations of metals were much lower in the heavy-textured Carswell soil, suggesting that soil texture was important. Also, the light-textured Newport soil had similar metal concentrations at 21 months to the Carswell soil even though the pH of Carswell was 1 unit lower, suggesting that texture was counteracting the effects of lower pH.

d) The effects of incubation and pH

Zinc and nickel concentrations in solutions from all soils decreased between the 1 and 3 month samplings, when the sludge organic matter was decomposing most rapidly. Thereafter, zinc and nickel concentrations in Cottenham and Carswell soils increased with time, but did not in the other 2 soils. Copper concentrations in all soils were generally low and varied little. The increases in metal concentrations found here were associated with decreases in solution pH, which affect zinc and nickel but do not affect copper until much lower pH values (1).

e) Metal speciation in soil solutions

The proportion of metals present as the free ion M^{2+} increased with time in the soils treated with metal-loaded sludges, with the exception of copper in the Newport and Upton soils (Table IV). This increase was due to a combination of the falling soluble organic C concentrations which caused a decrease in the number of potential complexing sites, and the falling pH which resulted in an increase in competition for complexing sites by H^{+}.

There were large differences between the proportions of the 3 metals present in solution as the free ion M^{2+} in the pH range studied. Copper was present predominantly as complexed species, zinc largely as the free ion and nickel as a mixture of both free and complexed species. Although copper concentrations varied little over the pH range found in the present experiment, copper free ion concentrations were very pH dependent, and this may influence plant availability of the soluble copper.

Table IV. Proportion of free ion present in soil solution displaced from replicates treated with metal-loaded sludges.

Period since sludge application Months	pH	Zn^{2+} %	Cu^{2+} %	Ni^{2+} %
		COTTENHAM		
1	7.6	28	0	7
3	7.4	20	0	32
7	6.8	78	7	36
11	6.5	77	2	52
16	6.2	87	14	45
21	6.4	81	16	70
		NEWPORT		
1	7.9	21	0	5
3	7.9	20	0	3
7	7.6	53	1	5
11	7.3	60	1	7
16	7.4	72	0	44
21	7.0	69	4	15
		CARSWELL		
1	7.5	27	2	3
3	7.1	70	0	100
7	7.2	75	16	14
11	6.8	85	2	28
16	6.7	92	25	61
23	6.3	100	22	77
		UPTON		
1	7.9	20	6	5
3	7.9	35	0	11
7	7.7	67	31	15
11	7.3	0	0	7
16	7.7	75	9	21
23	7.5	64	18	50

3.3. The effect of cropping on soil solution composition

The interpretation of data from cropped replicates is complicated by various factors. Firstly, the existence of fine rootlets may cause channelling in the soil and consequent dilution of displaced soil solution by the deionised water added to the soil surface. Secondly, some metal was removed by plants, and thirdly, the addition of fertilizers may have affected metal concentrations.

Soluble organic C and pH decreased with period of cropping, but the pH decreased more than in fallow treatments (Table V). The trend of increasing metal concentration with time found in the fallow pots was only observed for zinc and nickel in Cottenham soil in the cropped treatments. In this soil values were extremely high at 17 months in the red beet pots, presumably due to the low pH value (5.6), though in the other soils, metal levels were generally lower in cropped treatments than in fallow. It is probable that the trends in metal levels in the soil solution in cropped replicates were generally similar to those that occurred with fallow replicates, as the variation of organic C and pH with time was similar in both instances, but such variation was not observed in the data from the other 3 soils because of the factors described above.

3.4. Soil pH and content of DTPA extractable metals

The pH values of soil core samples removed from fallow pots at various times did not show the consistent trends with time that soil solution data did, and the range of values was smaller (Table VI). However, there were similarities between the two sets of data in that the pH values of the soil cores from Cottenham and Carswell were lower at 21 months than initially, whereas those for the other two soils were little changed.

The amount of metal extractable by DTPA from a given sludge treatment did not vary markedly from soil to soil, but was related to total metal contents, being about one third of the total soil values ($r = 0.87, 0.96, 0.70$ with $P < 0.001$ for Zn, Cu, and Ni respectively). The amount of metal extractable by DTPA at each sampling fluctuated somewhat. However, the extractable zinc levels were larger at 21 months than initially in all soils except Cottenham (Table VI), in contrast to the values found in soil solution (Table III). There were no discernible trends in copper or nickel data in Table VI, except that nickel may have increased slightly in Carswell soil.

3.5. Comparison of soil solution data, extractable metal and total metal contents with plant composition

The zinc and nickel contents of clover in two separate harvests were highly correlated with total metal, DTPA extractable metal and soil solution data from fallow pots sampled at the same time, when data from a single soil only was used, e.g. Cottenham (Table VII). The correlations for copper varied somewhat and were barely significant in harvest 3. However, when data from all 4 soils were bulked, there were marked differences between the three estimates of metal availability. For zinc and nickel soil solution concentrations were most closely related to plant composition, DTPA concentrations were less closely related, and total soil contents were the worst. For copper all three estimates were poor in harvest 3 but in harvest 10 soil solution

Table V. Composition of soil solution displaced from cropped treatments at various times after application of metal-loaded sewage sludge

Crop	Period since sludge application	pH	Soluble C	Sludge treatment					
				Zinc		Copper		Nickel	
				Low	High	Low	High	Low	High
	Months		mg C l^{-1}	μM	μM	μM	μM	μM	μM
				COTTENHAM					
Red	3	8.3	23	3.9	1.1	0.27	0.21	1.9	2.6
beet	12	6.8	21	8.3	9.9	0.85	1.5	1.4	1.9
	17	5.6	9	112.0	203.0	0.63	0.95	3.7	50
Clover	17	6.0	16	20.5	14.3	1.1	1.5	2.8	36
				NEWPORT					
Red	3	8.5	22	2.0	4.3	0.2	0.3	0.05	0.14
beet	12	7.2	12	3.3	4.4	1.6	2.1	0.82	1.2
	17	6.5	7	3.8	4.7	0.8	1.3	0.66	1.0
Clover	17	5.8	19	20.5	44	1.1	1.7	2.8	4.1
				CARSWELL					
Red	3	7.8	30	5.0	3.3	0.60	0.95	0.30	1.4
beet	12	6.6	30	2.4	15.0	0.79	1.07	0.39	0.7
	17	5.7	27	3.6	6.0	0.40	0.7	0.70	1.0
Clover	17	5.8	60	11.8	13.0	2.6	2.3	2.0	4.3
				UPTON					
Red	3	8.5	15	2.6	6.8	0.33	0.3	0.53	1.5
beet	12	7.7	18	0.66	2.6	0.40	0.95	0.30	0.4
	17	7.4	13	0.20	0.8	0.40	0.45	0.52	0.8
Clover	17	7.3	38	1.3	2.1	0.90	2.1	0.60	0.5

Table VI. Soil pH values and composition of DTPA extracts of soil samples in mg kg^{-1} soil, taken from fallow pots at various times.

Period since sludge application Months	pH	Sludge treatment					
		Zinc		Copper		Nickel	
		Low	High	Low	High	Low	High
				COTTENHAM			
0	6.3	65	99	23	29	13	15
1	6.5	47	92	16	26	5.3	9.9
7	6.3	46	105	14	34	5	12
16	5.49	80	160	29	42	13	23
21	5.95	54	85	14	29	5	12
				NEWPORT			
0	6.85	57	96	20	34	11	11
1	7.11	27	80	13	33	4	8
7	7.06	25	71	13	14	3	5
16	6.46	42	113	21	51	5	8
21	6.95	49	141	20	46	2	7
				CARSWELL			
0	6.55	52	118	19	49	8	16
1	6.33	48	75	19	33	7	15
7	6.28	63	85	16	32	9	16
16	5.63	98	134	37	66	11	18
21	5.98	88	130	31	53	13	24
				UPTON			
0	7.64	70	85	23	33	14	9
1	7.69	48	95	15	33	4	11
7	7.81	42	77	11	30	5	8
16	7.94	80	169	28	64	7	16
21	7.83	66	131	19	39	5	11

Table VII. Correlations between metal contents of clover crops and concentration of metals in soil solution, DTPA soil extracts, and the total metal content in the soil.

	Clover harvest 3 (6 months)			Clover harvest 10 (17 months)		
	Soil solution (7 months)	DTPA extractable	Total content	Soil solution (16 months)	DTPA extractable	Total content
COTTENHAM SOIL						
Zinc	0.94***	0.93***	0.97***	0.95***	0.96***	0.99***
Copper	0.68*	0.63*	0.62	0.89***	0.95***	0.73*
Nickel	0.99***	0.96***	0.93***	0.98***	0.95***	0.89***
ALL SOILS						
Zinc	0.94***	0.60***	0.44**	0.92***	0.54***	0.42**
Copper	0.50***	0.54***	0.53***	0.74***	0.50***	0.56***
Nickel	0.93***	0.67**	0.51*	0.86***	0.77***	0.51*

* $P<0.05$
** $P<0.01$
*** $P<0.001$

concentration was much better than the other 2.

Correlations between the limited solution data from clover-cropped pots and plant composition were less good than correlations between solution data from fallow pots and plant composition (r = 0.53 for clover pots and 0.92 for fallow pots for zinc over all soils at harvest 10) and were comparable to correlations between plant composition and DTPA-extractable metals.

3.6. Conclusions

a) The soil solution concentration of zinc, copper and nickel is affected by the total metal added to the soil, by the pH of the soil solution, and by the soil properties, such as texture and content of organic matter and free oxides.

b) The soil solution pH may decrease over a period of months after spreading sludge, and will affect the metal concentrations in the soil solution.

c) When data from several soils are bulked, the concentration of metals taken up by clover is more closely related to the metal concentrations in solutions displaced from fallow pots than it is to the concentrations of metals extracted by DTPA.

Acknowledgement - the authors thank the Water Research Centre, Stevenage and Severn-Trent Water authority for supplying the sludges.

REFERENCES

1. ADAMS, T.McM., SANDERS, J.R. (1983). Chemistry of zinc, copper and nickel from sewage sludge and sludge-treated soils. "Heavy metals in the environment", symposium proceedings, Heidelberg, C.E.P. consultants, Edinburgh.
2. ADAMS, T,McM., SANDERS, J.R. (1984a). The effects of pH on the release to solution of zinc, copper and nickel from metal-loaded sewage sludges. Environ. Pollut., series B, In press.
3. ADAMS, T.McM., SANDERS, J.R. (1984b). The effects of pH, soil type and incubation on the release to $CaCl_2$ solution of zinc, copper and nickel from 3 soils treated with metal-loaded sewage sludges. Unpublished.
4. LINDSAY. W.L., NORVELL, W.A. (1978). Development of DTPA soil test for Zn, Fe, Mn, Cu. Soil Science Society of America Journal 42, 421-428.
5. McBRIDE, M.B., BLASIAK, J.J. (1979). Zinc and copper solubility as a function of pH in an acid soil. Soil Science Society of America Journal 43, 866-870.
6. SANDERS, J.R. (1983). The effect of pH on the total and free ionic concentrations of manganese. zinc and cobalt in soil solutions. Journal of Soil Science, 34, 315-323.
7. WEBBER, M.D., KLOKE, A., TJELL, J.C. (1983). A review of current sludge use guidelines for the control of heavy metal contamination in soils. In "Processing and use of sewage sludge", symposium proceedings, Brighton 27-30 September 1983, ed. P.L'Hermite. D. Reidel, Dordrecht. (In press).

DISCUSSION

Chairman: R. LESCHBER

Rapporteurs: G.C. PORTEOUS, T.M. ADAMS, J.R. SAUNDERS

In answer to questions on the procedure for collecting soil solution samples, it was explained that all the experimental pots were maintained at 90% water capacity and small aliquots of de-ionised water were dosed very slowly prior to each sampling, the first 10ml drawn off being discarded as unrepresentative. Soil solution and core samples were taken from the blank soil pots without plants.

It was agreed that analysis of soil saturation extract would probably give similar results, but the method of periodic soil solution displacement had proved very useful in showing trends over a long period, such as the fall in pH, nitrates and sulphates, and the rapid initial decrease in soluble organic carbon in the sludged pots.

Interest was shown in the analytical methods used in the trials. Their use had been dictated by experience and equipment available at Rothamsted and they were not intended for routine use. Total metals in soils and crops had been determined by aqua regia digestion followed by inductively coupled plasma spectroscopy. Extractable metals were obtained with DTPA as work on weak extractants had not been planned at Rothamsted at the start of these trials. It was thought that there would be good correlation between metals extracted with sodium nitrate and calcium chloride and in soil solution. Electrical conductivity readings of soil solution were not taken.

The metal contents extracted from the soils by DTPA remained approximately constant, but those in the soil solution, except Ca and Mg, tended to increase with time. The soil solution concentrations of Zn, Cu and Ni correlated well with the amounts added in sludge and uptake by clover. There was general agreement that the free metal ion was an important factor in uptake by crops, and for specific soils and crops the ratios of metals in soil solution and crops were approximately constant. It was particularly important in bioavailability investigations that all relevant conditions should be stated.

Discussion of experimental error concentrated on whether or not replicate samples should be pooled, as other work had shown 10% variation in results from replicates. In this work all replicates had been analysed separately and statistical analyses of the results was still in progress.

It was considered that the soil solution investigation had made a valuable contribution to the assessment of Zn, Cu and Ni bioavailability in contaminated soils and supplemented the work done with saline extractants and complexing agents. However, various methods were in use in different laboratories for obtaining soil solutions and further investigations were needed to define the most suitable procedure. Accurate determination of the low levels of Cd likely to be found in soil solution presented a considerable analytical problem.

(Contributors to discussion : De Root, Hani, Herms, Kick, Muntau, Sauerbeck, Tjell, Webber, Worthington).

FACTORS INFLUENCING HEAVY METAL AVAILABILITY IN FIELD EXPERIMENTS WITH SEWAGE SLUDGES

C. JUSTE and P. SOLDA
Station d'Agronomie, INRA, Domaine de la Grande-Ferrade,
33140 PONT-DE-LA-MAYE (France)

Summary

Two long term field experiments were initiated in 1974 and 1976 in the same sandy soil located at proximity of Bordeaux (France). The purpose of this experiment was to measure the effects of heavy sewage sludge applications on the yield and the major and minor element content of a continuous maize crop.

1. MATERIAL AND METHODS

In the experiment initiated in 1974. the applied sludge (N.I) was a liquid anaerobically digested sludge with a low level of heavy metals.

In the experiment initiated in 1976, the applied sludge (N.II) was a solid anaerobically digested sludge (Porteous process) with a high level of heavy metals (Cd and Ni) arising from a battery factory.

The following treatments were realized (5 replicates) :

1. Control plots with only mineral fertilizers applied;
2. Annual application rates of farmyard manure equivalent to 10 t/ha of dry matter;
3. Annual application rates of sewage sludge equivalent to 10 t/ha of dry matter;
4. 100 t/ha of dry matter sewage sludge applied every two years.

The test-plant was a Maize grain with the same cultivar (INRA 260) utilized since the start of the experiment.

The total heavy metal amounts incorporated in the soils from 1974 or 1976 to 1983 through the sewage sludge applications and the soil concentration (measured in 1981) are indicated in table 1.

The heavy metal content of plants was measured, each year, at the following stages :

- a. Twelfth leaf fully developed stage (sixth leaf analysis);
- b. Tasseling stage (ear leaf analysis);
- c. Maturity stage (grain analysis).

2. RESULTS

Considering that the heavy metal content of plants reflects the true availability of heavy metals, this study provided the following observations.

1. Effect of the plant stage development on heavy metal availability (Table 1)

The metal content in the young plants was always more important than in the old plants (tasseling and maturity stages). The decrease of the heavy metal concentration in old tissues was presumably caused by the root colonization of the poorly metal enriched subsoil. This decrease might also indicate a dilution effect.

2. Influence of soil temperature (table 3)

In 1980, the Ni concentration of maize leaves increased drastically and, in the same time, plant yields decreased. The maize plants were stunted and exhibited strong phosphorus deficiency signs with low levels of this element in the plant tissues. Noticeable differences in soil temperature were observed in 1980 among the other years : during June, the soil temperature was indeed three degrees lesser than in the years before or after 1980.

3. Interaction effects of the various metals present in the sewage sludge (table 4)

In the years of the N.I sewage sludge application, the manganese content of maize was always drastically higher than the Mn content of the maize growing in the control plots. At the contrary, application of N.II sewage sludge decreased Mn content of maize below the level measured in control plots. This behaviour is difficult to explain because the Mn content of all sewage sludge treated plots was higher than the Mn content of control plots.

In 1980, the Ni uptake by maize sharply increased but, in the same time, Cd uptake was significantly decreased : presumably, low levels of soil temperature improved nickel uptake but not cadmium uptake; in these conditions, the decrease in the Cd uptake may be the result from the improvement of the physiological nickel availability.

4. Effect of plant species (table 5)

In 1979, between two successive maize crops, the field receiving highly metal polluted sewage sludge was cultivated with a lettuce crop. The data showed no effect of this treatment on the lettuce yield but a higher Cd accumulation was observed in the maize leaves whereas nickel was preferentially concentrated in lettuce. Thus, for a similar availability of metal in the soil, it was confirmed that some metal uptake was strongly related to plant species (1).

3. CONCLUSION

This study suggested that the true heavy metal availability was strongly controlled by physiological mechanisms. Therefore, chemical extractant utilization to predict metal uptake would be made with caution, namely in field experiments where the climatic and the soil factors in the same way as the complex composition of sewage sludge and the plant species are sharply involved in metal uptake.

REFERENCES

(1) JUSTE C. and SOLDA P., 1980 - Effect of heavy application of sewage sludge with high content of Cd and Ni on maize and lettuce crops. 2nd European Symposium on Characterization, treatment and use of sewage sludge, Vienne (Austria), 699-706.

Table 1

Amounts of heavy metal added to soils and aqua regia extractable content of metals measured in 1980 (0-20 cm depth)

Heavy metal	Sewage sludge * (N.I)		Sewage sludge ** (N.II)	
	Heavy metal added (kg/ha)	Heavy metal soil content (ppm)	Heavy metal added (kg/ha)	Heavy metal soil content (ppm)
Fe	20 774	4 418	11 354	3 470
Mn	4 151	419	234	81
Cu	122	35	170	46
Zn	1 973	352	976	158
Cd	18	5	641	110
Ni	124	7.5	1 337	269
Pb	443	92	231	49
Cr	32	12	64	18
Cumulative amounts of sewage sludge applied to soil (t/ha dry matter basis)			* 500 (1974 to 1982) ** 300 (1976 to 1980)	

Table 2

Mean values * of heavy metal content of Maize at different stages of develop in the heavy sewage sludge treated plots (ppm, dry matter basis)

Sewage Sludge N.	Stage	Cu	Fe	Mn	Zn	Cd	Cr	Ni	Pb
I	12 leaf development stage (sixth leaf)	18	115	169	244	3.1	0.4	1.0	4.0
	Tasseling stage (ear leaf)	16	156	147	179	1.0	0.4	1.3	4.5
	Maturity stage (grain)	0.5	24	8	41	-	-	-	-
II	12 leaf development stage (sixth leaf)	11	127	62	58	53	0.6	9.3	2.5
	Tasseling stage (ear leaf)	13	166	66	47	28	0.6	4.6	3.9
	Maturity stage (grain)	1	23	3	28	0.3	-	2.5	-

* Mean values measured from 1974 to 1983 (sewage sludge N. I)
Mean values measured from 1976 to 1983 (sewage sludge N. II)

Table 3

Fluctuations of Cd and Ni contents of maize ear leaf from 1974 to 1983 in the heavy sewage sludge treated plots

Cumulative amounts of sewage sludge added to soil (t/ha D.M. basis)	Year	Cadmium content (ppm D.M. basis)	Nickel content (ppm D.M. basis)	Maize grain yield (t/ha D.M. basis)	% of control (yield)
100	1976	17	4	10.2	97
100	1977	21	4	9.2	86
200	1978	29	3	8.0	79
200	1979	31	3.8	7.9	80
300	1980*	17	7.6	5.1	49
300	1981	53	5.1	8.3	91
300	1982	51	4.5	8.7	83
300	1983	30	4.6	9.5	98

* In 1980, the soil temperature at the beginning of the active development stage of maize (June) was 2° C lower than the corresponding temperature measured before or after this year.

Table 4

Manganese control on Maize (ear leaf)

cultivated in heavy sewage sludge treated plots and in control plots

Year	Mn control of ear leaf (ppm, dry matter)					
	Sewage sludge N. I			Sewage sludge N. II		
	Sewage treated plot	Control plot	% of control	Sewage treated plot	Control plot	% of control
1974	-	-	-			
1975	109	186	59			
1976	228	208	110	307	257	119
1977	51	98	52	67	113	59
1978	204	84	243	36	71	51
1979	52	55	95	34	124	27
1980	225	69	326	18	56	32
1981	60	103	58	22	95	23
1982	278	76	366	23	69	33
1983	112	106	106	23	106	22

Table 5

Differential uptake of Cd and Ni by Lettuce and Maize crops cultivated in the same field experiment (1979)

	Lettuce crop			Maize crop (*)		
	Cd content (ppm D.M.)	Ni content (ppm D.M.)	Cd/Ni	Cd content (ppm D.M.)	Ni content (ppm D.M.)	Cd/Ni
Control plots	5.9	1.7	3.5	4.7	2.1	2.2
Farmyard manure treated plots	4.3	1.4	3.1	1.8	1.9	0.9
Sewage sludge treated plots (10 t/ha/year)	27.2	15.4	1.8	47.4	3.6	13.2
Sewage sludge treated plots (100 t/ha/ 2 years)	42.6	46.4	0.9	70.8	8.5	18.6

(*) Maize crop : sixth leaf at 12 leaf development stage.

CONCLUSIONS

R.D. DAVIS and R. LESCHBER

1. Total concentrations of metals in sludge-treated soils are essential to define the extent of contamination.

2. Bioavailable concentrations of metals in soils are required to assess the significance of contamination in terms of effects on crops and the food chain.

3. Neutral salt extractants provide the most promising chemical method for assessing bioavailability. Both 0.1M sodium nitrate (Hani's paper) and 0.1M calcium chloride (Sauerbeck's paper) have been shown in conjunction with pot trials to work well for this purpose. In order to extend their application these extractants must be further evaluated using data from field experiments. Calcium chloride is a particularly promising extractant because it removes more metal and poses less of an analytical problem. If possible, the method should be extended to include nickel and copper in addition to cadmium and zinc.

4. Progressive acidification (Kiekens' paper) and sequential extraction techniques (Forstner's paper) are more arduous than single-step procedures and are therefore less suitable for routine use. As research methods they offer a way of characterising forms of metal in soil at least on an empirical basis, and of evaluating the long-term bioavailability of metals.

5. The soil solution (Adams' paper) is the interface between the plant root and the soil and therefore the concentration of metal in soil solution is likely to be closely related to bioavailability. Investigation of the chemical processes controlling the entry of metals into soil solution should provide a theoretical basis for predicting the effect of soil conditions on bioavailability.

6. Statistical errors associated with the sampling and analysis of soils and crops need to be defined and then minimised.

7. Improvements in analytical techniques for the determination of concentrations and forms of metals in soil extracts and plant digests must continue to be sought after.

LIST OF PARTICIPANTS

ADAMS, T.M.
Rothamsted Experimental Station
Harpenden
- HERTS
UNITED KINGDOM

AICHBERGER, K.
Landw.-chem. Bundesanstalt
Wieningerstrasse 8
4025 - LINZ
AUSTRIA

AMMANN, P.
Institut du Génie de l'Environnement
Génie Biologique

1260 - NYON-CHANGINS
SWITZERLAND

BECK, H.
Max-von-Pettenkofer Institute
Bundesgesundheitsamt
Postfach
1000 - BERLIN 33
GERMANY

BUSCH, C.
Agrikulturchemisches Institut der
Universität Bonn
Meckenheimer Allee 176
5300 - BONN 1
GERMANY

COPPOLA, S.
University of Naples
Istituto di Microbiologia Agraria
80055 - PORTICI
ITALY

CROSSMANN,
Josef-Könings-Institut
Mevinghoff 40
Postfach 5925
4400 - MUNSTER
GERMANY

DAVIS, R.D.
Water Research Centre
Medmenham Laboratory
Henley Road
Medmenham
P.O. Box 16
SL7 2HD - MARLOW, BUCKS
UNITED KINGDOM

DE GROOT, A.J.
Institut for Soil Fertility
Oosterweg 92
- HAREN
THE NETHERLANDS

DUVOORT-VAN-ENGERS, L.E.
Stichting Verwijdering Afvalstoffen
Natriumweg 7
P.B. 184
3800 AD - AMERSFOORT
THE NETHERLANDS

EADES, J.F.
The Agricultural Institute
Oak Park Research Centre
- CARLOW
IRELAND

FÖRSTER, C.
Niedersächsisches Landesamt
für Bodenforschung
Bodentechnologisches Institut
Friedrich-Missler-Str. 46-50
2800 - BREMEN
GERMANY

FÖRSTNER, U.
Arbeitsbereich Umweltschutztechnik
Tech. Univ. Hamburg-harburg
Eissendorfer Str. 38
2100 - HAMBURG 90
GERMANY

GOMEZ, A.
I.N.R.A.
Station d'Agronomie
Centre de Recherches de Bordeaux
Domaine de la Grande Ferrade
33140 - PONT DE LA MAYE
FRANCE

HAENI, H.
Forschungsanstalt für Agriculturchemie
und Umwelthygiene
Schwarzenburgstr. 155
3097 - LIEBEFELD (BE)
SWITZERLAND

HARDWICK, D.C.
Ministry of Agriculture, Fisheries
and Food
Great Westminster House
Horse Ferry Road
SW1P 2AE - LONDON
UNITED KINGDOM

HERMS, U.
Niedersächsisches Landesamt für
Bodenforschung
Bodentechnologisches Institut
Friedrich-Missler-Str. 46-50
2800 - BREMEN
GERMANY

JUSTE, Ch.
I.N.R.A. Station d'Agronomie
Centre de Recherches de Bordeaux
Domaine de la Grande Ferrade
33140 - PONT DE LA MAYE
FRANCE

KENSTEN, M.
Niedersächsisches Landesamt
Für Bodenforschung
Bodentechnologisches Institut
Friedrich-Missler-Str. 46-50
2800 - BREMEN
GERMANY

KICK, H.
Agrikulturchemisches Institut der
Universität Bonn
Meckenheimer Allee 176
5300 - BONN 1
GERMANY

KIEKENS, E.
Technical College for Chemistry, Textile
and Agriculture - C.T.L.
Voskenslaan 270
9000 - GENT
BELGIUM

KUNTZE, R.
Niedersächsisches Landesamt
für Bodenforschung
Bodentechnologisches Institut
Friedrich-Missler-Str. 46-50
2800 - BREMEN
GERMANY

LESCHBER, R.
Institute for Water-, Soil and
Air-Hygiene
Federal Health Office
Corrensplatz 1
1000 - BERLIN 33
GERMANY

LINDGREN, O.
The National Swedish Environmental
Protection Board
Box 1302
171 20 - SOLNA
SWEDEN

MINDERHOUD, A.
Rijksinstituut voor de Volksgezondheid
Anthonie van Leeuwenhoeklaan 9
Postbus 1
3720 BA - BILTHOVEN
THE NETHERLANDS

MUNTAU,
JRC
Ispra Establishment
 - ISPRA (Varese)
ITALY

PLUQUET, H.
Niedersächsisches Landesamt
für Bodenforschung
Bodentechnologisches Institut
Friedrich-Missler-Str. 46-50
2800 - BREMEN
GERMANY

PORTEOUS, G.C.
Department of the Environment
Romney House
43, Marsham Street
SW1 3PY - LONDON
UNITED KINGDOM

RELLFELD, H.
Josef-Könings-Institut
Mevinghoff 40
Postfach 5925
4400 - MUNSTER
GERMANY

SALT, C.
Entro Esserungswerke Berlin
Abt. Laboratorien
Freiheit 15-17
1000 - BERLIN
GERMANY

SANDERS, J.R.
Rothamsted Experimental Station
Harpenden
 - HERTS
UNITED KINGDOM

SAUERBECK, D.
Inst. f. Pflanzenernährung u.
Bodenkunde der FAL
Bundesallee 50
3300 - BRAUNSCHWEIG
GERMANY

SCHELTINGA, H.M.J.
Staatstoezicht op de Volksgezondheid
Pels Rijckenstraat 1
P.O. Box 9013
6800 DR - ARNHEM
THE NETHERLANDS

SEIFERT, D.
Josef-Könings-Institut
Mevinghoff 40
Postfach 5925
4400 - MUNSTER
GERMANY

STYPEREK, P.
Inst. f. Pflanzenernährung u.
Bodekunde der FAL
Bundesallee 50
3300 - BRAUNSCHWEIG
GERMANY

TARADELLAS, J.
Institut du Génie de l'Environnement
Génie Biologique
1260 - NYON-CHANGINS
SWITZERLAND